Future Trends in Soft Materials with Advanced Light Sources: Interdisciplinary of Bio- and Synthetic- Materials and Industrial Transferring

Journal of Physics: Conference Series Volume 272

Hyogo, Japan
1 – 3 September 2010

ISBN: 978-1-61782-301-5
ISSN: 1742-6588

Printed from e-media with permission by:

Curran Associates, Inc.
57 Morehouse Lane
Red Hook, NY 12571

Some format issues inherent in the e-media version may also appear in this print version.

Printed by Curran Associates, Inc. (2011)

Institute of Physics
Dirac House, Temple Back
Bristol BS1 6BE UK

Phone: 44 1 17 929 7481
Fax: 44 1 17 920 0979

techtracking@iop.org

Additional copies of this publication are available from:

Curran Associates, Inc.
57 Morehouse Lane
Red Hook, NY 12571 USA
Phone: 845-758-0400
Fax: 845-758-2634
Email: curran@proceedings.com
Web: www.proceedings.com

TABLE OF CONTENTS

PREFACE

This special issue contains peer-reviewed invited and contributed papers that were presented at The International Symposium on 'Future Trend in Soft Material Research with Advanced Light Source: Interdisciplinary of Bio- & Synthetic-Materials and Industrial Transferring', which was held in SPring-8, Japan, on September 1–3, 2010. Advanced light sources including neutron and synchrotron are becoming increasingly critical to the study of soft materials. This cutting-edge analytical tool is expected to lead to the creation of new materials with revolutionary properties and functions. At SPring-8, a new beam line dedicated to soft materials has now been launched as one of the most powerful X-rays for scattering and diffraction. Additionally, the next-generation light source, XFEL (X-ray Free Electron Laser), facilities are currently being developed in several locations. In the near future, femto-second and coherent X-ray sources will be available in soft material research and should reveal the various new aspects of advanced soft material research and technology.

On the occasion of the third fiscal year of the CREST (project leader: Kazuo Sakurai) and ERATO (project leader: Atsushi Takahara) projects, we organized this international symposium in order to accelerate the discussion among global-level researchers working on next-generation synchrotron radiation science, biophysics and supramolecular science, modern surface science in soft materials, and industrial applications of neutron and synchrotron radiation sources. In this symposium 21 oral presentations, including 8 invited speakers from abroad, and 40 poster presentations from USA, France, Korea, Taiwan, and Japan were presented during the three day symposium. The symposium chairs reviewed the poster presentations by young scientists, and eight young researchers received the Award for Best Poster Presentation.

We sincerely hope that these proceedings will be beneficial in future applications of advanced light sources to soft materials science and technology, not only to our ERATO and CREST projects, but also to the research of all the participants, broadening our scientific horizons.

Kazuo Sakurai & Atsushi Takahara
Symposium Chairs

Symposium Organization and Committee

Supported by:
Japan Science and Technology Agency (JST)
Japan Synchrotron Radiation Research Institute (JASRI)

Co-sponsored by:
Society of Japan Polymer Science
Japanese Society of Synchrotron Radiation Research
Advanced Softmaterial Beamline Consortium

Symposium Chairs:
Atsushi Takahara (Kyushu University, JST, ERATO)
Kazuo Sakurai (Univ. Kitakyushu, JST, CREST)

Organizing Committee:
Yoshiyuki Amemiya (The Univ. of Tokyo, JST, CREST)
Naoto Yagi (JASRI, JST, CREST)
Masaki Takata (JASRI)
Isamu Akiba (Univ. Kitakyushu, JST, CREST)
Yuya Shinohara (The Univ. of Tokyo, JST, CREST)
Taiki Hoshino (Kyushu University, JST, ERATO)
Jun-ichi Imuta (Kyushu University, JST, ERATO)
Moriya Kikuchi (Kyushu University, JST, ERATO)
Motoyasu Kobayashi (Kyushu University, JST, ERATO)

Group photograph

Lecture meeting

Journal of Physics: Conference Series **272** (2011) 012001 doi:10.1088/1742-6596/272/1/012001

Synchrotron Small Angle X-Ray Scattering Quantitatively Detects Angstrom Level Changes in the Average Radius of Taxol-Stabilized Microtubules Decorated with the Microtubule-Associated-Protein Tau

Myung Chul Choi[1-3‡], Uri Raviv[1-3§], Youli Li[4], Herbert P Miller[3], Daniel J Needleman[1-3†], Mahn Won Kim[5], Leslie Wilson[3,6], Stuart C Feinstein[3,6] and Cyrus R Safinya[1-3]

[1]Materials, [2]Physics, [3]Molecular, Cellular, and Developmental Biology Departments, University of California Santa Barbara, CA 93106
[4]Materials Research Laboratory, University of California Santa Barbara, CA 93106
[5]Department of Physics, KAIST, Daejeon 305-701, S. Korea
[6]Neuroscience Research Institute, University of California Santa Barbara, CA 93106

[‡]Current address: Department of Brain and Bioengineering, KAIST, Daejeon 305-701, S. Korea [§]Current address: The Chemistry Institute, Hebrew University of Jerusalem, Givat Ram 91904, Israel; [†]Current address: School of Engineering and Applied Science; Molecular and Cellular Biology; and FAS Center for Systems Biology, Harvard University, Cambridge, MA 02138

E-mail: myungchul.choi@gmail.com or safinya@mrl.ucsb.edu

Abstract. With the emerging proteomics era the scientific community is beginning the daunting task of understanding the structures and functions of a large number of self-assembling proteins. Here, our study was concerned with the effect of the microtubule-associated-protein (MAP) tau on the assembled structure of taxol-stabilized microtubules. Significantly, the synchrotron small angle x-ray scattering (SAXS) technique is able to quantitatively detect angstrom level changes in the average diameter of the microtubules modeled as a simple hollow nanotube with a fixed wall thickness. We show that the electrostatic binding of MAP tau isoforms to taxol-stabilized MTs leads to a controlled increase in the average radius of microtubules with increasing coverage of tau on the MT surface. The increase in the average diameter results from an increase in the distribution of protofilament numbers in MTs upon binding of MAP tau.

1. Introduction

Microtubules (MTs) are nanometer scale hollow cylinders derived from the eukaryotic cell cytoskeleton [1,2]. The hollow cylinders are comprised of globular tubulin subunits, which align end-to-end to form linear protofilaments (figure 1). The lateral interactions between protofilaments stabilize the tubular wall. In cells, MTs and their assembled structures are key components in a broad range of functions from providing tracks for the transport of cargo (organelles, vesicles containing neurotransmitter precursors, etc.), to forming the spindle structure in cell division, and to imparting mechanical stability in the long axons of neurons [1-4]. From a biomolecular materials perspective MTs may also be considered as model nanotubes with a well-defined inner (≈ 15 nm) and outer (≈ 25 nm) diameter, which can be fine-tuned depending on the solution conditions. Nanotubes and bundles of nanotubes are of high interest because of their potential technological, biomedical, and biotechnological applications [5-7].

Microtubules are highly dynamical biopolymers. They are usually found in two types of populations: Unstable MTs rapidly shrink (depolymerize) and grow (polymerize) (an important cellular property, which enables, for example, MTs to "search" and "capture" chromosomes in dividing cells during mitosis), whereas, in non-replicating mature neurons MTs are longer-lived in axons and dendrites. Distinct members of microtubule-associated-proteins (MAPs) control these two populations through MT-MAP interactions, although the precise manner in which MAPs affect MT structure and dynamics and MT-MT interactions is not well understood. The current model is that in neuronal axons, different isoforms of the microtubule-associated-protein (MAP) tau (shown in Figure 2) enhance tubulin assembly and regulate MT dynamics although much of the details remain unknown.

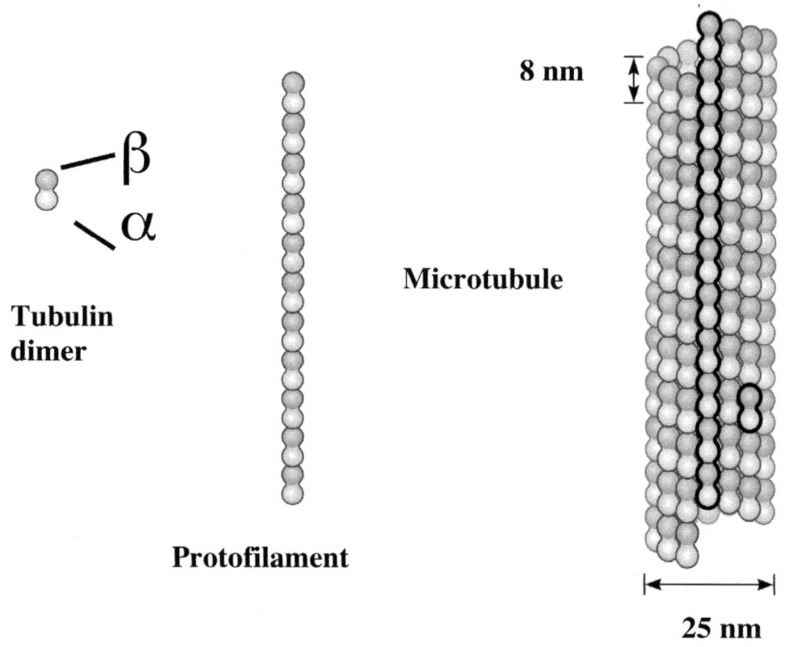

Figure 1. The α/β tubulin dimer is the building block of protofilaments. Protofilaments interact laterally in stabilizing the tubular wall of the microtubule. The cartoon of the microtubule (redrawn from [3]) highlights a tubulin dimer and a protofilament.

The very high interest in elucidating the precise nature of the interactions between MAP tau and MTs is the well known fact that altered tau-MT interactions, which changes the balance in the fraction of tau bound to the MT surface versus soluble tau, leads to MT depolymerization and tau tangles. This behavior, in turn, has been implicated in a large number of neurodegenerative diseases, including fronto-temporal dementia with Parkinsonism linked to chromosome 17 (FTDP-17), late stage Alzheimer's disease, Pick's disease, and supra-nuclear palsy [8-12].

2. Experimental Results and Discussion

Human MAP tau exists in the form of six isoforms, those containing four C-terminal "imperfect" repeats (constituting the cationic MT binding domains) labeled 4R-tau, and those with three such repeats (3R-tau) (figure 2) [13]. Each family contains three distinct alternatively spliced

Journal of Physics: Conference Series **272** (2011) 012001 doi:10.1088/1742-6596/272/1/012001

isoforms, yielding N-terminal short (S), medium (M), and long (L) projection domains (figure 2). The six tau sioforms are thus labeled: 4RS, 4RM, 4RL, 3RS, 3RM, and 3RL.

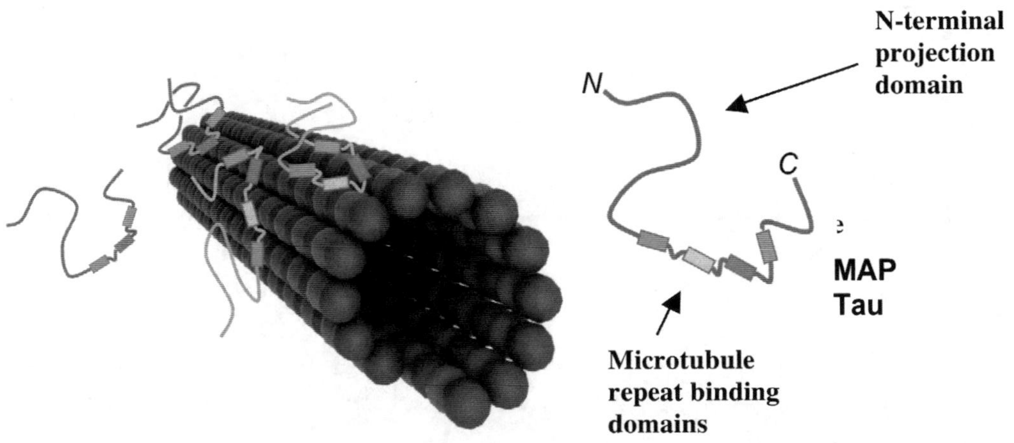

Figure 2. Schematic of a microtubule (MT) decorated with adsorbed MAP tau isoforms and two tau proteins in solution. MTs are nanotubes (outer diameter ≈ 25 nm) and are overall negatively charged. In their biologically active state tau isoforms are unstructured polypeptides, which possess either 3 "imperfect" repeat (3R) or 4 "imperfect" repeat (4R) cationic binding domains (colored rectangular boxes). The isoforms also differ in their N-terminal projection domain by possessing either zero, one, or two 29 amino acids, thereby generating short (S-), medium (M-), or long (L-) isoforms.

Synchrotron small angle x-ray scattering (SAXS) was used to quantitatively study the assembly structure of taxol-stabilized microtubules mixed with the six human MAP tau isoforms [14]. Figure 3 (Left) shows SAXS profiles of taxol-stabilized microtubules in the presence of the 4RL-tau isoform as a function of (Φ) defined to be the molar ratio of tau to tubulin-dimer in the reaction mixture. The observed intensity oscillations as a function of scattering wave vector q may be quantitatively described in terms of the MT form factor (F_{MT}) corresponding to the fourier transform of the MT electron density (figure 3, Middle). The MT was modeled as a simple nanotube with a uniform wall electron density (figure 3, Middle). The SAXS intensity I(q) is therefore proportional to $|F_{MT}|^2$ averaged over all directions in q-space [14].

The SAXS profiles are found to quantitatively fit the form factor model of a microtubule as can be seen from the colored solid lines going through the data in figure 3 (Left). We used a MT wall thickness of 49Å and a mean electron density of 0.07817e/Å^3 consistent with previously published papers [15-17]. Therefore the inner MT radius ($<R_{in}^{MT}>$) was the only fitting parameter. This model was also used previously to describe SAXS data of microtubules in the absence of MAP tau [18,19]. Consistent with earlier work we subtracted a background, which was a polynomial equation that goes through the minima observed in the SAXS profiles [14]. Figure 3 (Left) shows that the MT form factor is shifted with the addition of tau towards lower q values. This is consistent with an increase in the inner diameter of the microtubule.

The results of $<R_{in}^{MT}>$ obtained from the fits of the SAXS profiles to the data are shown in figure 3 (Right). The data is plotted both as a function of Φ and ϕ. While the former is the tau to tubulin-dimer molar ratio in the reaction mixture, the latter corresponds to the same ratio but only counting the tau molecules, which are bound to the MT surface (i.e. without the soluble tau fraction). The increase in the MT radius results from an increase in the average number of protofilaments in MTs (i.e. due to a shift in the distribution of MT protofilament numbers $<N_{pf}>$)

also shown in figure 3 (Right). Significantly, $<R_{in}^{MT}>$ was observed to rapidly increase for $0 < \phi < 0.2$ and saturate for ϕ between 0.2 and 0.5. Thus, a local shape distortion of the tubulin dimer upon tau binding, at coverages much less than a monolayer, is spread collectively over many dimers on the scale of protofilaments. Our findings imply that MAP tau regulates the shape of protofilaments and thus the spontaneous curvature of microtubules leading to changes in the curvature.

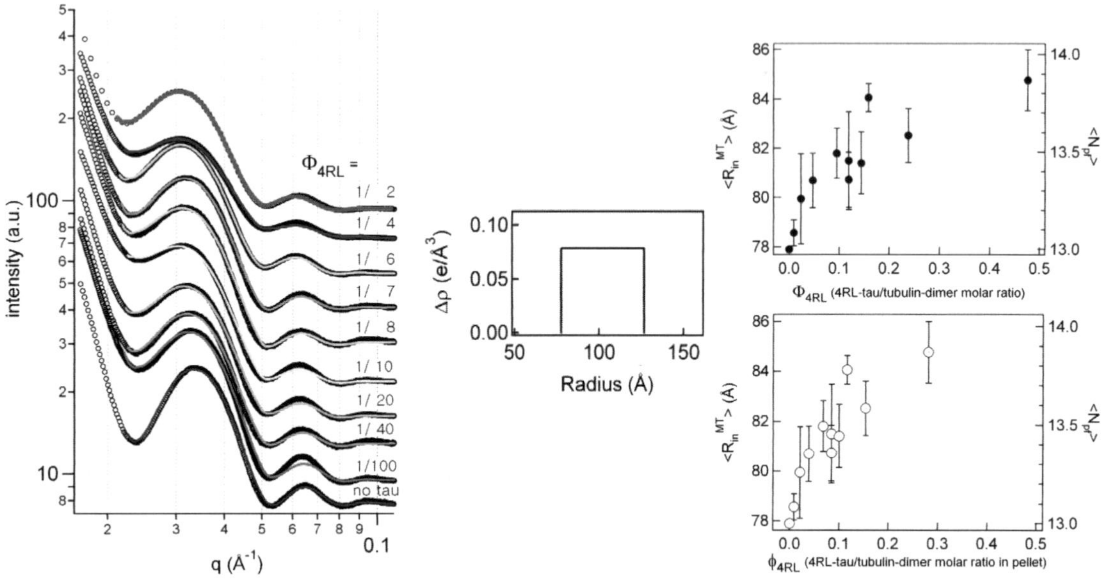

Figure 3. SAXS data showing that MAP tau regulates the mean radius of taxol-stabilized MTs, or equivalently, the microtubule (MT) protofilament number (N_{pf}). **LEFT:** Synchrotron SAXS profiles of MTs with bound 4RS tau as a function of tau to tubulin-dimer molar ratio in the reaction mixture (Φ). With increasing tau, the SAXS profiles shift to lower q implying an increase in the MT radius. Colored lines are results of fits of the data to a model consisting of a hollow MT nanotube with the inner radius $<R_{in}^{MT}>$ being the fitting parameter. **MIDDLE:** The electron density profile of the wall of the microtubule used to calculate the form factor of the microtubule. **RIGHT:** The inner radius $<R_{in}^{MT}>$ for 4RL-tau isoforms plotted versus Φ (TOP) and ϕ (BOTTOM), where ϕ is the molar ratio of tau adsorbed to the MT surface to the tubulin dimer (i.e. it is proportional to the number of tau molecules bound to the MT surface per unit area). The radial size of MTs increases as a function of increasing tau. The increase in $<R_{in}^{MT}>$ is a reflection of the shift (towards larger protofilament numbers) in the distribution of protofilaments in MTs with increasing tau binding [adapted from 14]. The SAXS profiles for Φ = 1/2, 1/6, 1/10, 1/20, 1/100, and no tau are adapted from [14]. The SAXS profiles for Φ = 1/4, 1/7, 1/8, 1/40, which show the same trend as the previously published data reported in [14] are new SAXS data.

3. Concluding Remarks

Although electron microscopy remains a very important probe of the structure of microtubules, Synchrotron SAXS, which yields statistically averaged structures, is a highly quantitative probe of the parameters characterizing individual microtubules such as the wall thickness and the electron density of the microtubule. Thus, the combination of these complementary techniques can quantitatively yield the structure of both isolated MTs and their assemblies [14,18,19]. Aside from the biophysical interest in the behavior of microtubules and associated proteins, microtubules may also be considered as model nanotubes and thus of high interest in applications. For example, Nanotubes are indispensable components in the development of future miniaturized materials. They have applications in diverse areas from uses as chemical, drug, and gene encapsulation vehicles, to biosensors and circuitry components. Thus, an area of intense current research in nanoscience is in elucidating the key parameters, which control the self-assembling properties of nanometer scale tubes. In this paper we described how MAP tau isoforms regulate the average radial size of taxol-stabilized MTs at the ångstrom scale. In particular, we found that tau regulates the distribution of protofilament numbers in MTs as reflected in the observed increase in the average radius $<R_{in}^{MT}>$ of MTs with increasing coverage of MTs by tau. The results may impact biomolecular materials applications requiring radial size controlled nanotubes.

Acknowledgments
CRS, YL, MCC, UR, DJN acknowledge support by the U. S. DOE BES DE-FG02-06ER46314 (plasmid preparation and protein binding and self-assembly) and the U. S. NSF DMR-0803103 (protein phase behavior). SCF was supported by the U. S. NIH NS35010 and LW and HPM were supported by the U. S. NIH NS13560. MCC further acknowledges support in part from the Korean Foundation Grant KRF-2005-2214-C00202. UR was also supported by the Human Frontiers Science Program Organization (CDA 0059/2006) and the Israel Science Foundation (grant number 351/08). DJN was further supported by a Human Frontiers Program Grant number RGP0034/2010 and the U. S. NSF grants DBI -0959721, PHY-0847188, and the Harvard MRSEC DMR-0820484. UR and DJN were jointly supported by the US-Israel Bi-national Science Foundation (grant number 2009-279). MK was supported by Korea Health 21 R&D Project MOHW and the WCU (World Class University) program through the National Research Foundation of Korea funded by the Ministry of Education, Science and Technology No. R33-2008-000-10163-0. The Stanford Synchrotron Radiation Laboratory, where the x-ray scattering work was performed, is supported by the U.S. Department of Energy. CRS acknowledges useful discussions with KAIST Faculty where he has a World Class University Visiting Professor of Physics appointment.

References
[1] Alberts, B et al *Molecular Biology of The Cell* (Garland 4th edition New York 2002).
[2] Desai A and Mitchison T 1997 *Annu. Rev. Cell Dev. Biol.* **13** 83.
[3] Bray D *Cell Movements: From Molecules to Motility*, (Garland 2nd edition N Y 2001).
[4] Kandel E R, Schwartz J H, Jessell T M *Principles of Neural Science*, (McGraw Hill 4th edition Singapore 2000)
[5] Martin C R 1994 *Science* **266** 1961.
[6] September issue on nanotechnology 2001 *Scientific American*, *285*
[7] Bong D T, Clark T D, Granja J R, Ghadiri M R 2001 *Angew. Chem. Int. Ed.* **40** 988.
[8] Brunden K R, Trojanowski J Q, Lee V M Y 2009 *Nature Rev. Drug Discov.* **8** 783.
[9] Kosik K S, Joachim C L, Selkoe D J 1986 *Proc. Natl. Acad. Sci. U.S.A.* **83** 4044.
[10] Hutton M, Lendon C L, Rizzu, P, Baker M et al 1998 *Nature* **393** 702.
[11] Clark L N, Poorka P Wszolek, Z Geschwind D H et al 1998 *Proc. Natl. Acad. Sci. U.S.A.* **95** 13103.
[12] Spillantini M G, Murrell J R, Goedert M, Farlow M R, Klug A, Ghetti B 1998 *Proc. Natl. Acad. Sci. U.S.A.* **95** 7737.
[13] Amos L A and Schlieper D 2005 *Adv. Protein Chem.* **71** 257.
[14] Choi M C, Raviv U, Miller H P, Gaylord M R, Kiris E Ventimiglia D, Needleman D J, Kim M W, Wilson L, Feinstein S C, Safinya C R 2009 *Biophysical Journal* **97** 519.
[15] Li H, DeRosier D J, Nickolson W V, Nogales E & Downing K H 2002 *Structure* **10** 1317.
[16] Lee J C, Frigon R P and Timasheff S N 1973 *J. Biol. Chem.* **248** 7253.
[17] Amos L A 1995 *Trends in Cell Biology* **5** 48.
[18] Needleman D J, Ojeda-Lopez M A, Raviv U, Miller H P, Wilson L and Safinya, C R 2004 *Proc. Natl. Acad. Sci. USA* **101** 16099.

[19] Raviv U, Needleman D J, Li Y, Miller H P, Wilson L and Safinya, C R 2005 *Proc. Natl. Acad. Sci. USA* **102** 11167.

Journal of Physics: Conference Series **272** (2011) 012002 doi:10.1088/1742-6596/272/1/012002

Beads-on-String Structure of the Electrostatic Complex of DNA with a High-Generation PAMAM Dendrimer

Chun-Jen Su[1], Chun-Yu Chen[1], Hsin-Lung Chen[2*] and Viktor A. Ivanov[3]
[1]National Synchrotron Radiation Research Center, Hsin-Chu 300, Taiwan [2]Department of Chemical Engineering, National Tsing Hua University, Hsinchu, Taiwan. [3]Physics Department, Moscow State University, Moscow 119991, Russia

E-mail: hlchen@che.nthu.edu.tw

Abstract. The electrostatic complexes of polyanionic DNA with cationic dendrimer have been considered as a potential non-viral vector for gene delivery and a model system for understanding DNA-histone interaction. Although it is believed that the gene transfection efficiency may be influenced by the structure of the complex, the supramolecular structure of DNA-dendirmer complexes and its dependence on various system parameters such as dendrimer generation number, charge density, charge ratio and ionic strength are not well resolved. In this study, we investigate the structure of the complex of DNA with polyamidoamine (PAMAM) dendrimer of generation nine (G9) by means of synchrotron small angle X-ray scattering (SAXS). It is found that DNA is always able to wrap around the dendrimer to yield the beads-on-string structure irrespective of the charge density of the dendrimer. The effect of charge density on the persistence length of the chromatin-like fiber thus formed and the pitch length of the DNA superheix wrapping around the dendrimer are elucidated from the calculation of the SAXS profiles based on beads-on-string structure models. The first section in your paper

1. Introduction

The electrostatic complexes of polyanionic DNA with various cationic agents, including lipids, macrocations, polyelectrolytes and amphiphilic block copolymers, have received much attention in recent years due to the effort in developing non-viral vectors for gene therapy. [1] The complexation is driven mainly by the electrostatic attraction between DNA and the cationic species coupled with the entropic gain from counterion release and it usually results in significant aggregation of DNA chains, leading to the formation of submicrometer-sized particles. [2,3] Two levels of the structure can hence be defined for the complexes: (1) the "colloidal level" characterized by the topological feature (e.g. the shape and size) and the surface charge of the particles at the length scale of several hundred nm or above; (2) the "supramolecular level" characterized by the organization of DNA chains and the cationic agent within the particles at the characteristic length scale of several nm. It is believed that the gene transfection efficiency is influenced by the structure of the complex; [4,5] consequently, resolving the structures at both levels and the strategy for tuning them by various parameters such as charge ratio, ionic strength, pH and temperature have been regarded as an important fundamental task for the realization of effective non-viral gene vector.

The present study concerns the supramolecular structure of the complexes of DNA with cationic dendrimer (called "dendriplexes") in pure water. Dendrimer is a class of hyperbranched macromolecule composing of layers of monomer units irradiating from a central core. [6] Each

Journal of Physics: Conference Series **272** (2011) 012002 doi:10.1088/1742-6596/272/1/012002

complete grafting cycle is called a "generation" (denoted by Gn with n being the generation number). The dendrimers possessing amine groups at the surface and/or the interior can be protonated to controlled level under acidic aqueous environment. The macrocations thus formed have been considered as the carriers for macromolecular drug and gene delivery. [7]

Based on the above discussion, it is suggestive that the complexes of DNA with high-generation dendrimers (e.g. G7 ~ G9) may exhibit the beads-on-string structure, while those with low-generation dendrimers (e.g. G2 and G3) may show the ordered columnar mesophase. The more complicated systems will hence be the complexes with dendrimers of intermediate generations (e.g. G4 ~ G6). In this case, neither the DNA bending energy nor the electrostatic interaction may completely dominate the structure formation; therefore, the dendriplexes may exhibit interesting structure transformation with respect to the interplay between these two factors prescribed by the charge density of the dendrimer, the charge ratio, salt concentration, etc. Moreover, some structures intermediate between the columnar mesophase and the beads-on-string structure may be formed. Using synchrotron small angle X-ray scattering (SAXS), we have shown recently that, depending on the charge density of the dendrimer prescribed by its degree of protonation (dp), the complexes of DNA with PAMAM G4 dendrimer exhibited three distinct nanostructures characterized by different degrees of DNA bending. [17] At $dp < 0.3$ the dendriplex displayed a square columnar phase, while the beads-on-structure with DNA wrapping around each dendrimer tightly by ca. 1.4 turns was formed at $dp \geq 0.6$. An intermediate structure called "hexagonally-packed DNA superhelices" was identified at $0.3 \leq dp \leq 0.5$, where the DNA chains organized in a hexagonal lattice twisted moderately to enhance the charge matching with the dendrimer. The observed structural transition with respect to the increase of dendrimer dp was in accord with the increasing weighting of electrostatic attraction over DNA bending energy.

In this paper, we investigate the structure of the complex of DNA with PAMAM G9 dendrimer to examine if DNA can always wrap around this high-generation dendrimer with various charge densities due to lower DNA bending energy cost. Some previous attempts have been made to reveal the wrapping of DNA around dendrimer. For example, Ottaviani et al. used electron paramagnetic resonance (EPR) spectroscopy to study the interactions between dendrimers and DNA and nitroxide-labeled PAMAM dendrimers. [18,19] DNA chain was found to wrap around G7 dendrimer. Similarly, Chen et al. investigated the binding of dendrimers to ctDNA by fluorescence titration of Ethidium bromide (EB) with a premixed solution of DNA and various amount of dendrimers. [20] The DNA complexation with PAMAM G7 dendrimer was found to be analogous to the DNA-histone interaction as DNA could wrap around the dendrimer. These spectroscopic studies could reveal the local interaction of DNA with dendrimer; however, the knowledge of the ''beads-on-string'' structure at a larger length scale remains largely unknown. Previous AFM studies have observed the globule particles formed by DNA-dendrimer complex, indicating that the DNA chain adsorbed on the surface of dendirmer could be bent due to strong electrostatic interaction. [21] Although detailed internal structure of these particles was not observed by AFM, Ritort et. al. used force-extension curves (FECs) to show similar unfolding and refolding force as well as similar compaction ratios to those of chromatin fibers. [22]

In this study, synchrotron SAXS is employed to gain insight into the structure information ranging from the internal structure of the DNA-wrapped dendrimer particles to the global organization of these ''nucleosome-like'' particles. We will demonstrate that DNA chain is indeed able to wrap around G9 dendrimer tightly even for the dendrimers with extremely low charge density. It will also be shown that the global organization of the resultant nucleosome-like particles or the conformation of the "chromatin-like" fibers is influenced by the charge density of the dendrimer.

2. Experimentals

2.1 Materials. Linear Calf Thymus DNA was purchased from ICN and used without further purification. Its molecular weight was ca. 9.2 kbps. Ethylenediamine (EDA) core polyamidoamine G9 dendrimer was obtained from DNT as methanol solutions. After thorough drying, the solids were weighed and then redissolved in distilled water to produce a stock solution of 0.1% (w/w).The solutions were stored at 4 °C till use.

2.2 Complex Preparation. To complex with the polyanionic DNA, the amine groups in PAMAM dendrimer were first protonated by adding prescribed amount of 0.1 N HCl solution. The primary amine groups at the outer surface of the dendrimer tended to be protonated first because their basicity ($pK_a \cong 9.0$) is larger than that of the interior teriary amines ($pK_a \cong 5.8$). [23] Therefore, the PAMAM G9 dendrimers with different degrees of protonation (dp) were prepared by controlling the pH of the solution. The solution of the protonated dendrimer solution was then mixed with the aqueous solution containing prescribed amount of DNA to obtain the complex. The concentration of DNA aqueous solution was 2 mg/ml. The complexaton was usually manifested by visually observable precipitation.

2.3 Small Angle X-ray Scattering (SAXS) Measurements. The supramolecular structures of the complexes in pure water were probed SAXS at room temperature. The aqueous suspensions of the complexes were directly introduced into the sample cell comprising of two ultralene windows. The SAXS experiments were performed at the Endstation BL23A1 of the National Synchrotron Radiation Research Center (NSRRC), Taiwan. The energy of X-ray source and sample-to-detector distance were 14 keV and 2259 mm, respectively. The scattering signals were collected by MarCCD detector of 512x512 pixel resolution. For the in-house experiment, the wavelength of X-ray source and sample-to-detector distance were 0.154 nm (Cuα) and 65 cm (for the low-q configuration) or 23 cm (for the high-q configuration), respectively. The scattering intensity profile was output as the plot of the scattering intensity (I) vs the scattering vector, $q = (4\pi/\lambda)\sin(\theta/2)$ (θ = scattering angle), after corrections for sample transmission, empty cell transmission, empty cell scattering and the detector sensitivity.

3. Results and discussion

The charge density of the PAMAM G9 dendrimer is prescribed by its dp value which stands for the number fraction of protonated amine groups in the dendrimer. The nominal N/P ratio of the dendriplex, prescribed by the feed molar ratio of the amine groups (irrespective of whether they are protonated) to the phosphate groups of DNA, is fixed at 6/1.

Figure 1 displays the SAXS profiles of the dendriplexes with three dp values for the G9 dendrimer. The scattering patterns are very different from those associated with the ordered columnar phases exhibited by the dendriplexes with lower-generation dendrimers, [22] showing that DNA in the complexes does not organize to form columnar mesophases. For dp=0.02 and 0.1, the SAXS pattern is characterized by a shoulder (marked by "q_m") near 0.05 Å^{-1}, a small hump (marked by "*i=1*") at ca. 0.08 Å^{-1} and a large broad shoulder/peak near 0.14~0.15 Å^{-1}. When the dp is increased to 0.5, the SAXS profile displays a sharper primary peak and a small shoulder (marked by "*i=1*") appearing like a form factor peak. An additional peak (marked by "q_p") is identified at 0.24 Å^{-1}.

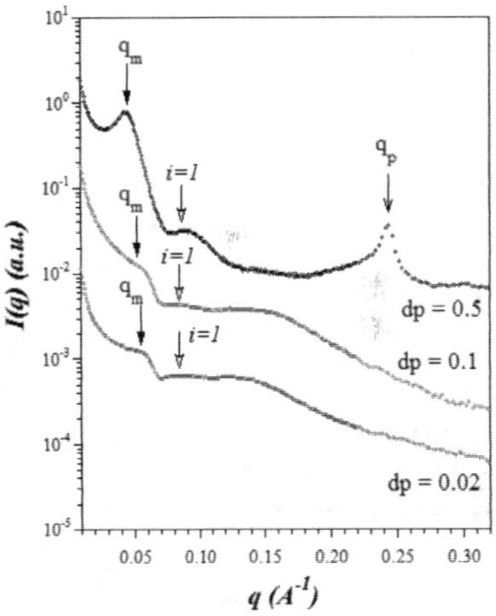

Figure 1. Room-temperature SAXS profiles of DNA-PAMAM G9 ddendrimer complexes as a function of dp. The nominal N/P ratio of the dendriplexes is 6/1.

Because the ordered columnar mesphase is not formed in DNA/G9 complexes, we resort to another possible structure, namely, the "beads-on-string" structure. A G9 dendrimer molecule (R = 59 Å) is much larger than a lower-generaiton dendrimer molecule, such as G2 (R = 12 Å) thereby rendering a much higher charge density in protonated G9 dendrimer (cf. 2048 positive charges on G9 vs. 16 positive charges on G2 for dp = 0.5). The strong electrostatic attraction between DNA and G9 dendrimer may induce the DNA chain to wrap tightly around the dendrimer molecule for effective charge matching. The energy cost in bending the DNA chain to wrap around a G9 dendrimer is also lower since the large molecular size reduces the surface curvature of the dendrimer.

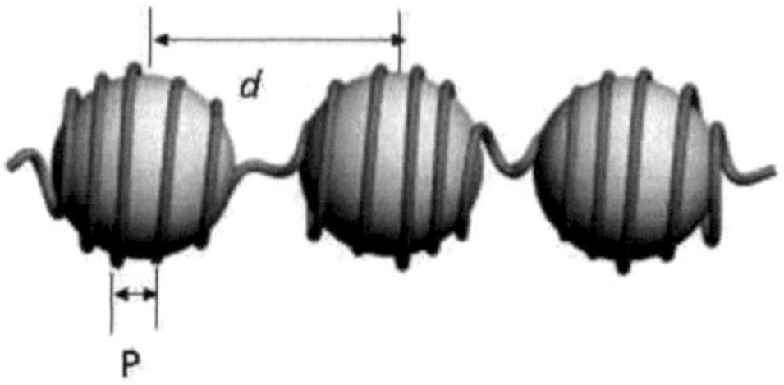

Figure 2. Schematic illustration of the chromatin-like beads-on-string structure; *P* and *d* correspond to the pitch length and the axial interparticle distance of the nucleosome-like particles, respectively.

It is known that the subunit of chromatin, called "nucleosome", consists of a 147-bp DNA wrapping around a 7 nm cationic octamer of histone protein. This kind of beads-on-string structure of chromatin or isolated nucleosome particles has been characterized by SAXS and SANS. We hence first consider the salient features of the scattering behavior of chromatin from the literature and then examine if the scattering behavior of DNA-G9 complex is analogous to that of chromatin.

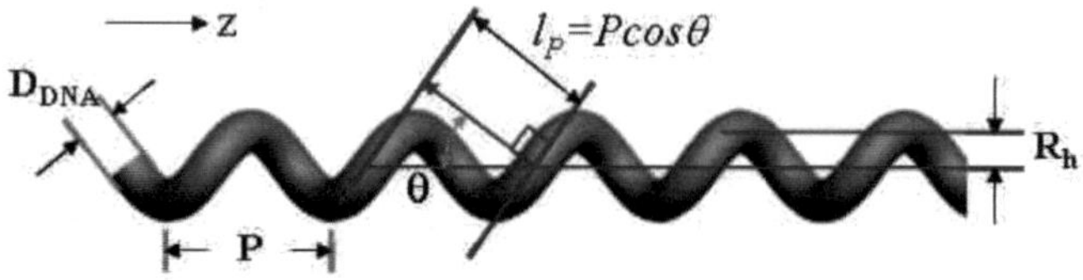

Figure 3. Definitions of the geometric parameters of a DNA superhelix used for model calculation. P and θ represent the pitch length and pitch angle, respectively. R_h is the radius of the superhelix defined as the radial distance between the centerline and central trace of the helix. l_P is the projection of P onto the normal of the helical segment.

Baldwin et al. reported a series of SANS profiles of chromatin dispersed in D_2O/H_2O mixture (the concentration of chromatin in the solution was 50 wt%) for contrast variation experimetn. [24] According to the authors, the SANS pattern collected at 10% D_2O approximates to the SAXS pattern. In this case, the scattering profile shows a low-q peak and two additional high-q peaks. The low-q peak with the equivalent Bragg spacing (d) of ca. 10.5 nm was attributed to the interparticle distance between the nucleosome particles constituting the chromatin fiber. The two high-q peaks were considered to stem from DNA component, where the peak corresponding to d = 5.5 nm was ascribed to the pitch of DNA wrapping around the histone protein. The observed SAXS patterns of dp 0.02 and 0.1 dendriplexes studied here are indeed quite similar to that SANS profile of chromatin, which strongly suggests that the DNA chain wraps around G9 dendrimer to yield the beads-on-string structure (as schematically illustrated in Figure 2) even at the dp as low as 0.02 (corresponding to 82 positive charges per dendrimer macrocation). Since the DNA used here has 9200 base pairs with the fully extended length of ca. 3 m, each DNA strand is able to wrap around a large number of G9 dendrimer, giving rise to the "chromatin-like fiber" composing of the "nucleosome-like particles".

Although the scattering result implies the formation of beads-on-string structure irrespective of dp, the fact that the SAXS profile depends on dp indicates that the internal structure of the chromatin-like fiber varies with dp. Following the work of Baldwin et al., [24] we consider the primary peak to stem from the spatial correlation of the nucleosome-like particles. The interparticle distance calculated from the peak position (q_m) *via* d = 2 /q_m is 12.6 nm. Considering that the diameters of DNA and dendrimer are 2.0 nm and 11.4 nm, respectively, the interparticle distance of the closely packed nucleosome-lile particles should be d = 11.4 + 2 x 2=15.4 nm. The observed d is however smaller than this value. We postulate that the primary scattering peaks in SAXS profiles are instead associated with the correlation of the dendrimer molecules along the fiber contour (i.e. z-axis; see Figure 2). The dendrimer molecules are closely spaced along the fiber axis as the interparticle distance is only slightly larger than the molecular diameter.

The chromatin-like fiber formed by the complex should possess certain persistence length. The fact that the primary peak is broad and weak in intensity at dp < 0.5 signals that the axial correlation is limited and hence the fiber has a relatively small persistence length (or more flexible). The primary peak becomes sharper and more intense when dp is increased to 0.5, implying that the complex fiber becomes stiff with large persistence length. In this case, the charge density of the dendrimer is very high (with 2048 charges on the surface), and a significant amount of positive charges remain unmatched even if DNA wraps around the dendrimer tightly. The complex is hence overcharged and the strong electrostatic repulsion between the overcharged nucleosome-like particles stiffens the fiber significantly. Moreover, the long-range correlation of the pitches due to persistent wrapping of DNA along the fiber axis leads to a relatively sharp and clear pitch peak located at ca. 0.24 Å $^{-1}$. The pitch length calculated from the peak position according to $P = 2\pi/q_P$ is 2.6 nm. This value is slightly larger than the diameter of DNA, showing that the DNA segments are closely spaced along the fiber contour to effectively match the positive charges on dendrimer at the cost of bending energy.

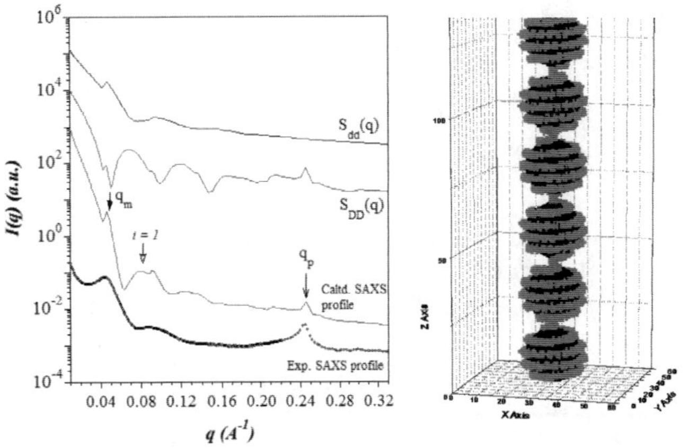

Figure 4. Comparison between the experimentally observed SAXS profile and the calculated SAXS pattern for the dp/0.5 dendriplex. The SAXS profile is calculated assuming a chromatin-like rod composing of 10 nucleosome-like particles placing in sequence with P=2.6 nm and d = 14 nm. The partial structure factors associated with DNA-DNA and dendrimer-dendrimer correlation, i.e. SDD(q) and Sdd(q), are also displayed. The figure on the right shows the actual picture of the chromatin-like segment generated for calculating the SAXS curves.

The formation of beads-on-string structure by DNA-G9 dendrimer complexes is formally verified by comparing the observed SAXS profiles with the calculated form factor of a chromatin-like fiber. We construct a chromatin-like rod formed by a DNA chain wrapping around a number of dendrimer macrocations placing along a fiber axis (i.e., z axis) with the axial interparticle distance of *d* (Figure 2). Each dendrimer is approximated by a sphere with internal monomer density fluctuations, [25] and the DNA superhelix is approximated by a uniform helical cylinder with a prescribed pitch length *P* and pitch angle, as shown in Figure 3. The radius of the superhelix (R_h) given by $R_h = P/(2\pi \tan\theta)$ is defined as the radial distance between the centerline and central trace of the helix ($R_h = 0$ for completely straightened DNA, see Figure 3). Therefore, the helical trace of the cylinder can be calculated from the following equations.[26]

$$x(z) = R_h \sin(\frac{2\pi z}{P} + \phi);\ y(z) = R_h \cos(\frac{2\pi z}{P} + \phi)$$

where ϕ is phase angle that prescribes the direction of the groove of the superhelix. The regular pitch of a helix can give rise to a scattering peak locating at $q_p = 2\pi/l_P = 2\pi/P\cos\theta$, where l_P is the projection of *P* onto the normal of the helical segment (see Figure 3). It can be shown that $q_p \approx 2\pi/l_P$ as long as $2\pi R_h$ is significantly larger than *P*. It is noted that the wrapping of DNA around the dendrimer is assumed to be tight; therefore, the value of R_h may vary with z.

After constructing the chromatin-like fiber with prescribed values of *P* and *d*, we divide the system into numerous volume elements (each with the size of 8 x 8 x 8 Å^3) and the partial structure factors associated with DNA-DNA correlation [$S_{DD}(q)$], dendrimer-dendrimer correlation [$S_{dd}(q)$] and

DNA-dendrimer correlation [$S_{Dd}(q)$] of the randomly oriented chromatin-like fiber are calculated by the Debye equation[27]

$$S_{nm}(q) = \frac{1}{N^2}\sum_{i=1}^{N}\sum_{j=1}^{N}\frac{\sin\left(q\left|\mathbf{r}_i - \mathbf{r}_j\right|\right)}{q\left|\mathbf{r}_i - \mathbf{r}_j\right|} \qquad (1)$$

where n and m stands for either DNA (D) or dendrimer (d) and $|\mathbf{r}_i - \mathbf{r}_j|$ is the distance between i and j volume element. $S_{dd}(q)$ is further corrected by adding an additional component arising from internal monomer density fluctuations.[25] The SAXS intensity is finally calculated from the three partial structure factors by

$$I(q) = \Delta\rho_D{}^2 S_{DD}(q) + 2\Delta\rho_D\Delta\rho_d S_{Dd}(q) + \Delta\rho_d{}^2 S_{dd}(q) \qquad (2)$$

where $\Delta\rho_D$ (= 5.7 x 10^{10} cm^2) and $\Delta\rho_d$ (= 2 x 10^{10} cm^2) is the scattering length density contrast of DNA and PAMAM dendrimer relative to water, respectively. It is noted that at present we have not been able to obtain the scattering profiles that quantitatively match the experimental results, as there are a large number of parameters needed to be considered, such as the distribution of SLD within the

dendrimer molecule, the possible size distribution of dendrimer arising from non-uniform protonation, the persistence length of the chromatin-like fiber, the wrapping mode of DNA, etc. Here we merely seek the structures that give rise to the scattering profiles resembling the experimental results, and from which we can identify the salient features of the beads-on-string structures formed at different degrees of protonation.

Figure 4 compares the experimental SAXS profile of the dendriplex with dp =0.5 with that calculated for a chromatin-like rod composing of 10 nucleosome-like particles placing in sequence with P = 2.6 nm and d = 14 nm. The partial structure factors associated with DNA-DNA and dendrimer-dendrimer correlations, i.e., $S_{DD}(q)$ and $S_{dd}(q)$, are also displayed in the figure. From the calculated SAXS profile, it can be seen that the scattering pattern at q < 0.12 $Å^{-1}$ is dominated by $S_{dd}(q)$, where the primary peak is associated with the interparticle distance between the dendrimers (or the nucleosome-like particles) along the z axis and the small hump near 0.08 $Å^{-1}$ is the first form factor maximum of the dendrimer. However, $S_{DD}(q)$ dominates the scattering pattern at q > 0.12 $Å^{-1}$. In this case, a clear pitch peak (q_P) becomes visible at 0.24 $Å^{-1}$. The intensity and breadth of this pitch peak is dependent on the number of nucleosome-like particles in the fiber assumed for the calculation. The peak drops in intensity and broadens with decreasing number of nuclesosome-like particles. Therefore, a chromatin-like fiber with larger persistence length should display a more clear pitch peak along with a sharper primary peak in the SAXS profile.

The close resemblance of calculated SAXS profile to the experimentally observed scattering pattern verifies the formation of beads-on-string structure by the dp/0.5 dendriplex. The calculation also confirms that the primary peak is associated with the axial correlation of the dendrimer (or nucleosome-like particle) in the

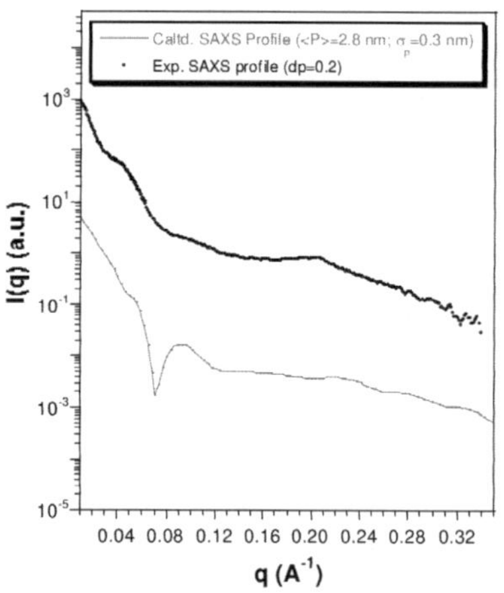

Figure 5. Comparison between the experimentally observed SAXS profile and the calculated SAXS pattern for the dp/0.2 dendriplex. The SAXS profile is calculated assuming a chromatin-like rod composing of four nucleosome-like particles with a Gaussian distribution of the pitch length (mean value = 2.8 nm and the variance =0.3 nm).

chromatin-like fiber and the peak marked by "*i = 1*" corresponds to the first form factor maximum of the dendrimer. The peak observed at 0.24 $Å^{-1}$ corresponds to the pitch peak found in the calculated profile; therefore, the clear pitch peak in the observed SAXS profile attests a rather large persistence length of the chromatin-like fiber and the pitch length of the DNA wrapping is 2.6 nm. Our present calculation is however not rigorous enough to provide an accurate estimate of the persistence length.

For dendriplexes with lower dp, no clear pitch peak is identified and the corresponding SAXS profiles in the high-q region display a broad shoulder. Moreover, the primary scattering peak is also broad and relatively weak. These features attest that the persistence length of the chromatin-like fiber is short. We found that the assumption of a monodisperse pitch length in the nucleosome-like particles cannot produce the SAXS profile showing a broad high-q shoulder. On the other hand, the assumption of polydisperse pitch length yields the SAXS pattern closely resembles the experimentally observed profiles. Figure 5 displays the experimental SAXS profile of the complex with dp = 0.2 and the SAXS profile calculated under the assumption of Gaussian distribution of the pitch length with the mean value of 2.8 nm and the variance of 0.3 nm. In this case, the intensity at $q > 0.05$ $Å^{-1}$ is obtained by summing the form factor profiles of the nucleosome-like particles with different pitch lengths according to the weighting prescribed by the Gaussian distribution function. The low-q intensity ($q <$ 0.05 $Å^{-1}$) is calculated for the chromatin-like rod composing of four nucleosome-like particles placed in sequence with the interparticle distance of 11.8 nm. It can be seen that the agreement is fairly good, thereby showing that the chromatin-like fiber formed by the dnedriplex with lower dp is more flexible (comparing to that associated with dp/0.5 dendriplex) and there exists a relatively clear distribution of pitch length of the DNA superhelix wrapping around the dendrimer.

4. Conclusion

We have revealed that DNA can wrap around PAMAM G9 dendrimer to yield the chromatin-like structure irrespective of the charge density of the dendrimer. The wrapping mode of the nucleosome-like particle and the global conformation of the chromatin-like fiber however depend on dendrimer charge density. At low dp (< 0.5), DNA can still wrap around the dendrimer tightly with a distribution of pitch length; the chromatin-like fiber thus formed has a smaller persistence length. At dp = 0.5, the DNA chain wraps around the dendrimer regularly and tightly with the pitch length of 2.6 nm. The resultant chromatin-like fiber is highly stiff due to strong electrostatic repulsion between the nucleosome-like particles.

5. Acknowledgemen

We acknowledge the financial support of the National Science Council of the Republic of China under grant No. NSC 97-2923-E-007-001.

6. References

[1] Niidome, T.; Huang, L. *Gene Therapy,* **2002**, 9, 1647
[2] Bloomfield, V. A. *Biopolymers,* **1991**, 31, 1471.
[3] Bloomfield, V. A. *Biopolymers,* **1997**, 44, 269.
[4] Koltover, I.; Salditt, T.; Rädler, J. O.; Safinya, C. R. *Science,* **1998**, 281, 78.
[5] Safinya,C. R. *Curr. Opin. Struc., Biol.,* **2001,** 11, 440.
[6] Tomaila, D. A.; Baker, H.; Dewald, J.; Hall, M.; Kallos, G.; Martin, S.; Roeck, J.; Ryder, J.; Smith, P. *Polym. J.*, **1985,** 17**,** 117.
[7] Esfand, R.; Tomaila, D. A. *Drug Discov Today,* **2001**, 6, 427.
[8] Prosa, T. J.; Bauer, B. J.; Amis, E. J. *Macromolecules,* **2001**, *34*, 4897.
[9] Tomalia, D. A.; Hall, M.; Hedstrand, D. M. *J. Am. Chem. Soc.,* **1987**, *109*, 1601.
[10] Bielinska, A. U.; Chen, C.; Johnson, J.; Baker, J. R., Jr. *Bioconjugate Chem.,* **1999**, *10*, 843.
[11] Stryer, L. *Biochemistry*; W.H. Freeman and Company: San Francisco, 1981.
[12] Kunze, K. K.; Netz, R. R. *Phys. Rev. Lett*., **2000**, 85, 4389.
[13] Kunze, K. K.; Netz, R. R. *Phys. Rev. E.*, **2002**, 66, 011918.

[14] Liu, Y. C.; Chen, H. L.; Su, C. J.; Liu, H. K.; Liu, W. L.; Jeng, U. *Macromolecules,* **2005**, 38, 9434.
[15] Su, C.J.; Chen, H.L.; Wei, M.C.; Peng, S.F.; Sung, H.W.; Ivanov, V.A. *Biomacromolecules*, **2009**, 10, 773.
[16] Evans, H. M.; Ahmad, A.; Ewert, K.; Pfohl, T.; Martin-Herranz, A.; Bruinsma, R. F.; Safinya, C. R. *Phys. Rev. Lett.,* **2003**, 91, 7, 075501-1.
[17] Chen, C.Y.; Su, C.J.; Peng, S.F.; Chen, H.L.; Sung, H.W. *Soft Matter*, **2010**, in press.
[18] Ottaviani, M. F.; Sacchi, B.; Turro, N. J.; Chen, W.; Jockusch, S.; Tomalia, D. A. *Macromolecules* ,**1999,** *32,* 2275.
[19] Ottaviani, M. F.; Furini, F.; Casini, A.; Turro, N. J.; Jockusch, S.; Tomalia, D. A.; Messori, L. *Macromolecules* **2000,** *33,* 7842.
[20] Chen, W.; Turro, N. J. ; Tomalia, D. A. *Langmuir* **2000,** *16,* 15.
[21] Mitra, A.; Imae, T. *Biomacromolecules*, **2004,** 5, 69.
[22] Ritort, F.; Mihardja, S.; Smith, S. B.; Bustamante, C. *Phys. Rev. Lett.* **2006**, 96, 118301.
[23] Cakara, D.; Kleimann, J.; Borkovec, M. *Macromolecules* **2003**, *36,* 4201.
[24] Baldwin, J. P.; Boseley, P. G..; Bradbury, M.; Ibel, K. *Nature* **1975,** 253, 245.
[25] Porcar, L.; Liu, Y.; Verduzco, R.; Hong, K.; Butker, P. D.; Magid, L. J.; Smith, G. S.; Chen, W. R. *J. Phys. Chem. B,* **2008**, 112, 14772.
[26] Hamley, I. W. *Macromolecules,* **2008,** 41, 8948.
[27] Debye, P. *Ann. Physik.,* **1915**, 46, 809.

X-ray Photon Correlation Spectroscopy Study on Dynamics of the Free Surface in Entangled Polystyrene Melt Films

Tadanori Koga[1,2,*], Chunhua Li[2], Maya K. Endoh[2], Suresh Narayanan[3], Laurence Lurio[4], Sunil K. Sinha[5]

[1]Chemical and Molecular Engineering Program, [2]Department of Materials Science & Engineering, Stony Brook University, Stony Brook, New York 11794-2275, [3]Advanced Photon Source, Argonne National Laboratory Argonne, IL, 60439, [4]Department of Physics, Northern Illinois University, DeKalb, IL, 60115, [5]Department of Physics, University of California San Diego, La Jolla, CA, 92093

E-mail:tkoga@notes.cc.sunysb.edu

Abstract. The dynamics of polymer chains near the surface of a melt and within thin films remains a subject of inquiry along with the nature of the glass transition in these systems. Recent studies show that the properties of the free surface region are crucial in determining the anomalous glass transition temperature (T_g) reduction of polymer thin films. In this study, by embedding "dilute" gold nanoparticles in polystyrene (PS) thin films as "markers", we could successfully probe the diffusive Brownian motion which tracks the local viscosity both at the free surface and within the rest of the single PS thin film far above bulk T_g. The technique used was X-ray photon correlation spectroscopy with resonance-enhanced X-rays that allows us to independently measure the motion in the regions of interest at the nanometer scale. We found the presence of the surface reduced viscosity layer in entangled PS thin films at $T \gg T_g$.

1. Introduction

It is well known that a nanoconfinement effect can play an important role in determining the mechanical and rheological properties of polymer thin films. With the advent of increasing miniaturization, lubricant films of this thickness have become commercially relevant and hence it is important to investigate the effect on a more fundamental level. So far, intensive experimental and theoretical studies have focused on the relationship between the glass transition temperature (T_g) and the chain dynamics in polymer thin films (see, for examples, Refs. [[1, 2]]). However, T_g values are often said to average over the entire film so that the information is an ensemble average including both a polymer/substrate interface and polymer/air interface. By using x-ray photon correlation spectroscopy with gold nanoparticles as "markers", we measured the Brownian motion of the gold particles at the free surface and the center of the polymer film independently. All of the gold relaxation functions were well described by a single exponential function and the enhanced surface dynamics was observed at all temperatures far above the T_g. These indicate that the origin of the enhanced surface mobility is not heterogeneous dynamics due to the glass transition. Further, we found that the gold dynamics tracked the viscosity of the polymer matrix and the viscosity of the free surface showed $\eta \propto M_w^{3.4}$, which is known as the power law variation for entangled polymers.

2. Experimental

Four different molecular weights of polystyrene (PS, Mw=123, 200, 290, and 400 kg/mol) having narrow polydispersities (M_w/M_n=1.02-1.05) were investigated. Thiol-functionalized (octadecanethiol ($C_{18}H_{37}SH$)) gold nanoparticles were prepared by using the one phase synthesis method developed[3]. The average radius of the core gold particles and the thickness of the $C_{18}H_{37}SH$ (shell) layer were approximated to be 15 Å and 13 Å, respectively. Solutions of PS were first prepared in toluene. The Au nanoparticles were then added to the PS solutions, which were further sonicated for 30 min. The weight percentages of the gold nanoparticles were fixed to 4%. PS/gold thin films were prepared by spin-casting onto HF etched Si substrates and annealed for 1 day in vacuum of 10^{-3} Torr at T=170 °C >T_g, in order to remove residual solvents and spin-induced stress. TEM experiments gave us the evidence that the individual Au particles were dispersed in the PS polymer matrix even after the XPCS experiments.

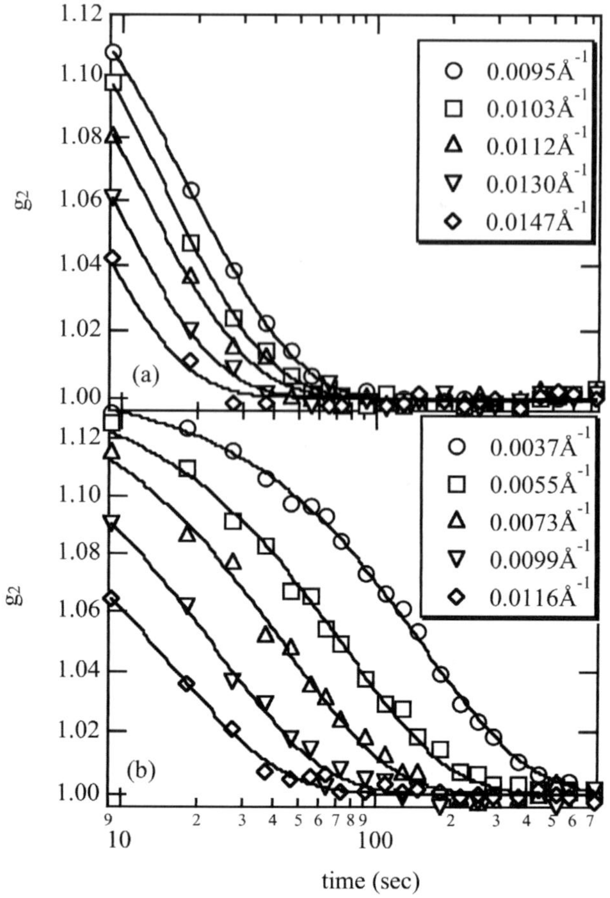

Fig.1 Measured g_2 at T =176 °C for the PS (M_w=123 kg/mol)/Au film at (a) the surface mode and (b) resonance mode. The solid lines are the best single exponential fits described in the text.

In order to study the surface dynamics alone, we used a "markers" X-ray photon correlation spectroscopy (XPCS) technique that serves a "fingerprint" of the motion of the gold nanoparticles in polymer matrices. Furthermore, by intensifying the probing electrical field in the regions of interest for XPCS experiments, we could successfully differentiate the nanoparticles motions at the surface from the rest of the film by using the following two modes: (i) The "surface-mode" with the incident angle (α_i = 0.15°) below the critical angle of the total external reflection of PS (0.16° with the x-ray energy of 7.5 keV used for the experiments). This allows us to study the surface area (about 90 Å depth) since the electric field intensity (EFI) decays exponentially into the film. (ii) The "resonance-mode" with the incident angle just above the critical angle where the first-resonance enhancement of the EFI in the polymer film takes place. Resonance-enhanced x-rays (REX), which are generated by the interference of x-ray plane waves reflected and refracted at the two interfaces of the film in air, are significantly intensified at the position close to the center of the film [4]. We found that the resonance enhanced x-ray intensity at the first resonance mode (i.e., at the incident angle (θ) of 0.168° for a 1280 Å film) is 53.5 times higher than that of the incident beam and the intensity near both interfaces can be eliminated[5]. It should be noted that the incident angle for the first resonance mode depends on the film thickness and second-order (θ=0.184°) and third-order (θ=0.200°) REXs can be also produced within the film. Consequently, a series of XPCS experiments at the two different modes allow direct investigation of the gold motion at the topmost surface area in a single thin film.

Journal of Physics: Conference Series **272** (2011) 012003 doi:10.1088/1742-6596/272/1/012003

The XPCS experiments were performed at the beam line 8-ID at the Advanced Photon Source (APS), Argonne National Laboratory. The details of the experiments have been described elsewhere[6]. XPCS measures the normalized intensity-intensity time autocorrelation function, $g_2(q_{//},t)$

$$g_2(q_{//},t) = \frac{<I(q_{//},t')I(q_{//},t+t')>}{<I(q_{//},t')>^2} \quad (1)$$

where $I\ (q_{//},\ t')$ refers to the scattering intensity at the in-plane wave vector transfer $q_{//}$ and at time t'. The brackets < > refer to averages over time t' and t is the delay time. $g_2(q_{//},t)$ is related to the normalized intermediate structure factor $[f(q_{//},t)]$ via $g_2(q_{//},t) = 1 + A[f(q_{//},t)]^2$, where A is the speckle contrast. g_2 at each $q_{//}$ could be fitted by using a single exponential function, $f(q_{//},t) = \exp(-t/\tau)$, where τ is the characteristic relaxation time of the function.

3. Results and Discussion

Fig. 1 shows representative g_2 functions at five different $q_{//}$ values at the surface mode (Fig. 1(a)) and the resonance mode (Fig. 1(b)) at 176°C. The thickness of the PS(M_w=123 kg/mol)/Au film was 1280 Å. The solid lines are best-fits of the exponentially decaying autocorrelation function to the experimental data. Fig. 2 shows the $q_{//}$ dependences of the τ values obtained at both modes at 176°C. From the figure we can see that all the $q_{//}$ dependences followed the power-law behavior of $\tau \propto q_{//}^{-2}$, indicating the diffusive motion of the gold nanoparticles in the polymer matrix. From the best-fits to the data (the solid lines) with the relationship of $\tau = 1/Dq_{//}^2$, where D is the diffusion constant of the gold particles, the D values were estimated to be 345Å^2/s for the surface mode and 235Å^2/s for the resonance mode, respectively. This indicates that the Brownian diffusive motion of the gold particles is approximately 50% faster at the surface than near the center of the film. It should be emphasized that the enhanced gold dynamics at the surface (~ 50%) could be observed for the PS/Au films (~130 nm thickness) regardless of temperatures (156 ≤ T ≤ 186 °C) and molecular weights of the polymers used in this study.

In order to investigate the origin of the enhanced gold dynamics, we calculated the theoretical D values using the Stokes-Einstein (S-E) law for diffusion of isolated particles in a viscous medium, i.e., $D = k_B T / 6\pi R\eta$, where k_B is the Boltzmann constant, T is the absolute temperature, R is the effective radius of the thiol-functionalized Au particles (28 Å), and η is the viscosity of the entire film. The averaged film viscosity for the PS/Au thin film was estimated to be 3.5×10^4 (Ns/m^2) at T=176 °C by using bilayer dewetting measurements[7]. Substituting the measured η value for the PS(M_w=123 kg/mol)/Au nanocomposite thin film into the S-E law gave us the calculated D values of 331Å^2/s, showing that the experimental D values obtained from the XPCS experiments are very close to the calculated D values. Therefore, we can conclude that the thiol-functionalized gold nanoparticles have

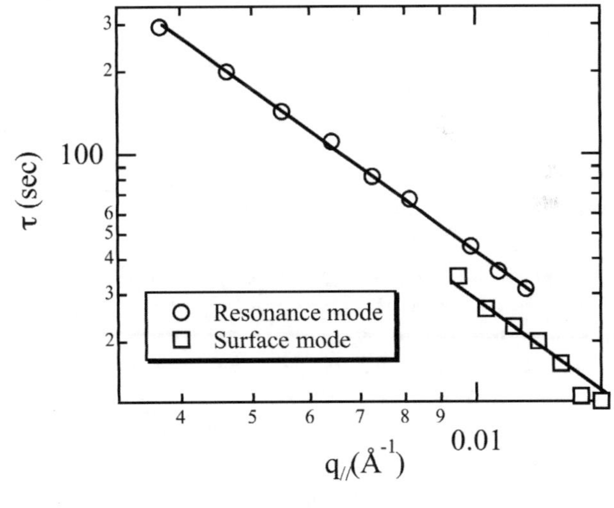

Fig.2 Log-log plot of τ vs. $q_{//}$ at the surface mode and resonance mode PS (M_w=123 kg/mol)/Au film at T=176 °C. The solid lines correspond to the best fits of the relationship ($\tau = 1/2Dq_{//}^2$) to the data.

a very weak interaction with the polymer such that the Brownian motion of the nanoparticles tracks the viscosity of the matrix. Consequently, the XPCS results clearly demonstrate that the enhanced Au dynamics is due to the presence of the reduced viscosity layer at the topmost surface of the single polymer film at T >> T_g.

We further investigated the local viscosity of the surface and center regions calculated by the S-E law. Interestingly, we found the well-known relationship of $\eta \propto M_w^{3.4}$, which is known as the power law variation for entangled polymers[8], in both the surface and resonance modes[5]. This finding indicates that the polymer chains are indeed entangled even at the surface region within the thickness of 90 Å and the chain dynamics is still governed by the pure reptation mechanism.

The next question to be clarified is whether this surface reduced viscosity layer is present regardless of the film thickness. Our preliminary XPCS experiments with different thicknesses of the PS/Au films gave us the evidence that the surface layer is perturbed when a distance from the substrate is close to 600 Å and the further reduction of the distance down to about 300 Å obliterates the surface layer to a great extent. We believe this is attributed to the presence of a "dead" layer which is strongly bound to the substrate and hence remains on the substrate even after thorough rinsing with a good solvent [9]. Further studies will provide a new insight into the mechanism of the nanoconfinement effect on the rheological property of polymer thin films.

4. Conclusion

In summary, by using the marker XPCS technique in conjunction with resonance-enhanced x-rays, we found the existence of the surface enhanced layer in the liquid polymer thin films with ~1300 Å thickness. The viscosity of the free surface layer was about 30 % lower than the rest of the film. Moreover, we found that the emergence of the surface enhanced layer was strongly related to the entanglement of the polymer chains, resulting in the well-known relationship of $\eta \propto M_w^{3.4}$ at the free surface of polymers. We believe that the methodology described here would open up a new avenue to explore *in-situ* and surface perturbation (such as oscillatory in the case of atomic force microscopy)-free dynamics at specific surfaces and interfaces of polymer films or membranes to be used for interdisciplinary materials research.

Acknowledgements

T.K. acknowledges the financial support provided by NSF CAREER award under funding No. CMMI-0846267 and ChemMatCARS, APS. Use of the Advanced Photon Source was supported by the U.S. DOE under Contract No. DE-AC02-06CH11357.

References

[1]Forrest J. A. 2002 *Eur. Phys. J. E* **8** 261
[2]Alcoutlabi M. and McKenna G. B. 2005 *J. Phys. Condens. Matter* **17** R461
[3]Yee C. K., Jordan R., Ulman A., White H., King A., Rafailovich M. H. and Sokolov J. C. 1999 *Langumuir* **15** 3486
[4]Wang J., Bedzyk M. and Caffrey M. 1992 *Science* **258** 775
[5]Koga T., Li C., Koo J., Endoh M. K., Rafailovich M. H., Narayanan S., Lee D. R., Lurio L. and Sinha S. K. 2010 *Phys. Rev. Lett.* **104** 066101
[6]Kim H., Ruhm A., Lurio L. B., Basu J. K., Lal J., Lumma D., Mochrie S. G. J. and Sinha S. K. 2003 *Phys. Rev. Lett.* **90** 068302
[7]Xavier J. H., Sharma S., Seo Y. S., Isseroff R., Koga T., White H., Ulman A., Shin K., Satija S. K., Sokolov J. C. and Rafailovich M. H. 2006 *Macromolecules* **39** 2972
[8]Doi M. and Edwards S. F., The Theory of Polymer Dynamics (Oxford Science, Oxford, 1986).
[9]Fujii Y., Yang Z., Leach J., Atarashi H., Tanaka K. and Tsui O. 2009 *Macromolecules* **42** 7418

Journal of Physics: Conference Series **272** (2011) 012004 doi:10.1088/1742-6596/272/1/012004

New opportunities for Anomalous Small-Angle X-Ray Scattering to characterize Charged Soft Matter Systems

M. Sztucki, E. Di Cola, and T. Narayanan

European Synchrotron Radiation Facility, B.P. 220, 38043 Grenoble, France.

E-mail: sztucki@esrf.eu

Abstract. This article presents a quantitative analysis of charged Soft Matter system using Anomalous Small-Angle X-Ray Scattering (ASAXS). Specific experimental requirements and data reduction are described. The simultaneous analysis of data sets recorded below *and* above the absorption edge provides valuable constraints in data modelling. The merit of the method is illustrated by an example involving cationic surfactant micelles. The counterion profile around the micelle is deduced in a self-consistent manner and the results reveal the strong condensation of counterions on to the micellar surface.

1. Introduction

Deeper understanding of charged soft matter systems is of direct relevance to many applied areas of research including biophysical processes [1]. Examples include charge stabilized colloids, polyelectrolytes, surfactant micelles and membranes, DNA, proteins, etc. An important issue in the detailed description of such systems is the spatial distribution of counterions around a macroion [2]. In this context, quantitative Anomalous Small-Angle X-ray Scattering (ASAXS) offers a direct method for the structural characterization of charged systems [3-8]. The spatial distribution of counterions can be deduced with high precision by tuning the energy in the vicinity of the absorption edge of the counterions [5,9]. This information is not readily accessible by conventional scattering techniques as the contributions of the counterions and the macroions superimpose and thus cannot be distinguished. However, the stability and radiation sensitivity of the specimens as well as relatively low concentration of anomalous species in typical soft matter systems put high demands on the instrumental setup. The technical requirements for accurate data recording and the theoretical background for quantitative analysis of anomalous small-angle X-ray scattering data have recently been summarized [10].

This article focuses on new opportunities offered by detecting and ideally discriminating the contribution of fluorescence, which plays an important role in the energy range close and above the absorption edges. It will be demonstrated that quantitative simultaneous fitting of the scattering curves recorded not only below [10], but also above the corresponding absorption edge is feasible and helps to determine the spatial distribution profile of the counterions with high accuracy. This is illustrated by a micellar surfactant system (Tetradecyl-trimethylamonium bromide, TTAB) near the bromine K-edge (13474eV).

2. Anomalous small-angle X-ray scattering

The scattered intensity, $I(q)$, from a suspension of centrosymmetric particles as a function of the modulus of the scattering vector, $q = \frac{4\pi}{\lambda}\sin(\frac{\theta}{2})$, (with θ being the scattering angle and λ the wavelength of the incident radiation), can be expressed as

$$I(q) = NP(q)S(q) \tag{1}$$

with N the number density of particles, $P(q)$ the single particle scattering function, and $S(q)$ the structure factor of interactions. The single particle scattering function, $P(q)$ is given by

$$P(q) = F(q)F^*(q) \tag{2}$$

$F(q)$ being the scattering amplitude, and $F^*(q)$ its complex conjugate. In the case of spherical particles

$$F(q) = 4\pi \int_0^\infty \Delta\rho(r) \frac{\sin(qr)}{qr} r^2 dr \tag{3}$$

with $\Delta\rho(r)$ the relative scattering length density, $\Delta\rho(r) = \rho(r) - \rho_m$, where $\rho(r)$ is the radial scattering length density profile and ρ_m is the scattering length density of the background.
For dilute non-interacting systems, $S(q) \approx 1$, but for concentrated samples a polydisperse effective hard-sphere structure factor will be used.

Anomalous small-angle X-ray scattering uses the energy dependence of the X-ray scattering factor f of a given element when the energy of the incident X-ray beam is varied near the absorption edge of that element [see Fig. 1(a) for bromine]. In this case, the number of excess electrons per ion immersed in a medium becomes a function of energy of the incident beam

$$\Delta f(E) = \Delta f_0 + f'(E) + if''(E) \tag{4}$$

where Δf_0 denotes the scattering contribution of a single ion, f' and f'' are the real and imaginary parts of f. The energy dependence of the radial scattering length density profile $\Delta\rho$ can be determined as

$$\Delta\rho(r,E) = \Delta\rho_0(r) + v(r)r_e[f'(E) + if''(E)] \tag{5}$$

with $\Delta\rho_0(r)$ being the non-resonant scattering length density, $v(r)$ being the spatial distribution of anomalous scattering units and $r_e = 2.818 \times 10^{-15}$m the classical electron radius. For further details see [10].
The total scattered intensity can be decomposed into three components,

$$I(q) = N\{F_0^2(q) + 2f'(E)F_0(q)v(q) + [f'^2(E) + f''^2(E)]v^2(q)\} \tag{6}$$

the energy independent normal SAXS $F_0^2(q)$, a cross-term involving the amplitudes of normal SAXS and the resonant scattering of the counterions, and the pure resonant scattering term (self-term) due to counterions $v^2(q)$. The Fourier transform of the self-term $v^2(q)$ can directly yield the spatial distribution $v(r)$ of the resonant atoms [7,9].

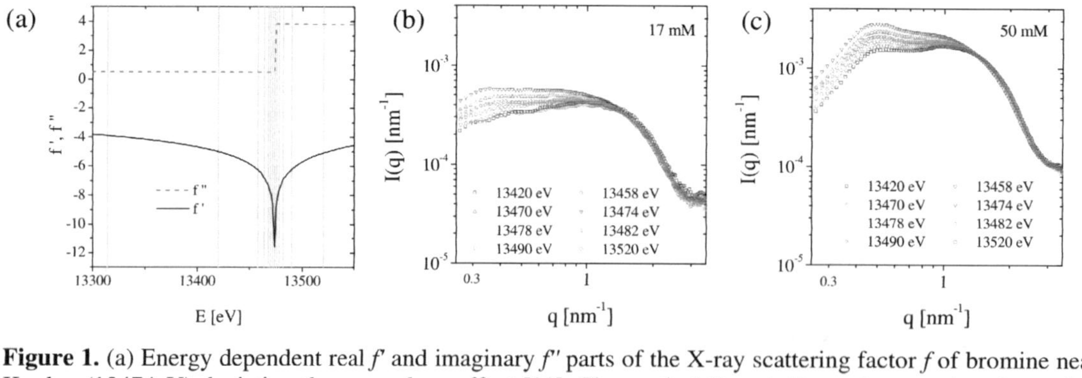

Figure 1. (a) Energy dependent real f' and imaginary f'' parts of the X-ray scattering factor f of bromine near its K-edge (13474eV) depicting the anomalous effect [11]. The vertical lines indicate the energies chosen in this study. (b) and (c) present a selection of the recorded ASAXS data at a sample concentration of 17mM and 50mM, respectively. The fluorescence background was subtracted from all data.

3. Experimental

Tetradecyl-trimethylamonium bromide [TTAB, $C_{14}H_{29}N(CH_3)_3Br$] is a well-known cationic surfactant, which forms micelles above the critical micellar concentration (cmc) of about 3.6mM. TTAB

(Aldrich 99%) solutions were prepared at different concentrations above the cmc: 10mM, 17mM and 50mM using MilliQ water.
ASAXS measurements were performed at the *High Brilliance Small-Angle X-ray Scattering* beamline ID02 at the European Synchrotron Radiation Facility, Grenoble (France). Measurements of the samples with 17mM and 50mM concentration were recorded at a sample to detector distance of 0.9m by using the FReLoN CCD detector [10]. The following relative energies (*E*-13474eV) near the bromine K-edge were used: -1000eV, -160eV, -54eV, -16eV, -10eV, -6eV, -4eV, -2eV, ±0eV, +2eV, +4eV, +8eV, +16eV, +46eV.
As fluorescence detection plays an important role in the energy range close to and above the absorption edges, further data were recorded using an energy-resolving detector (Pilatus300K from Dectris). The sample to detector distance for these measurements was 2.0m. The investigated sample had a concentration of 10mM. The detector was first operated with a standard energy threshold of 50% of the working energy (i.e. ~6700eV). A second measurement was performed using an adjustable threshold of 12500eV as the fluorescence emission of Bromine is about 11.9keV (Kα).

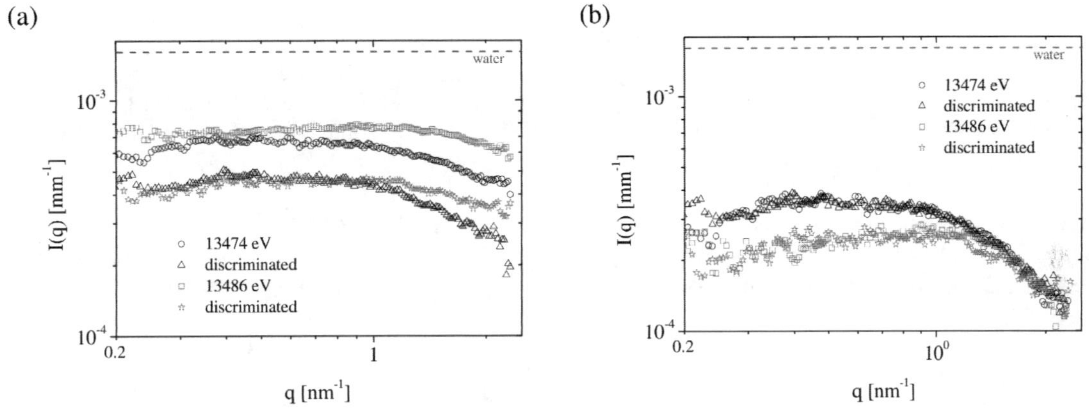

Figure 2. ASAXS measurement of a sample with 10mM concentration using an energy-resolving pixel detector (Pilatus300K). The scattering level of water is indicated to demonstrate the limit of detection. Fig. (a) shows the normalized scattering data recorded at two energies at and 12eV above the absorption edge of Br. Circles and squares represent data recorded with an energy threshold of 6700eV whereas the measurements marked by triangles and stars were recorded with maximum energy discrimination (threshold: 12500eV). In both cases a fluorescent background is detected which has to be subtracted. (b) After fluorescence correction, data recorded with and without energy discrimination are in good agreement.

Figure 2(a) shows normalized ASAXS data, which were recorded, without specific energy discrimination (threshold 6700eV) at the absorption edge of Br and 12eV above (circles and rectangles, respectively). Both curves show a considerable background of fluorescence (comparable to the fluorescence signal detected using the FReLoN CCD camera). By better energy discrimination with a threshold value of 12500eV, the florescence level can be considerably reduced (triangle and stars). Nevertheless, even the state-of-the-art 2D pixel detectors cannot provide the required level of discrimination to completely eliminate the fluorescence background. Figure 2(b) shows the additionally fluorescence-subtracted data for both threshold settings. Both datasets are in good agreement. The fluorescence signal is determined from the recorded scattering level at high q values.
The ASAXS data were analysed by simultaneously fitting multiple scattering curves recorded below and above the absorption edge of Bromine. The obtained results demonstrate that the fluorescence background can be well handled. Fitting ASAXS data recorded below and above the absorption edge allows achieving better reliability of the simulated data as the full achievable range of f' and f'' values are exploited. For data fitting, a spherical core double-shell model was used which is described in detail in [10]. Fig. 3(a) presents exemplarily 8 of the 14 fitted curves together with their best fits showing

good agreement. Fig. 3(b) illustrates the model which uses a polydisperse core of radius R with scattering contrast of the tetradecyl chains in the inner sphere (blue), followed by a first shell of closely packed trimethylammonium head groups (red, thickness t).

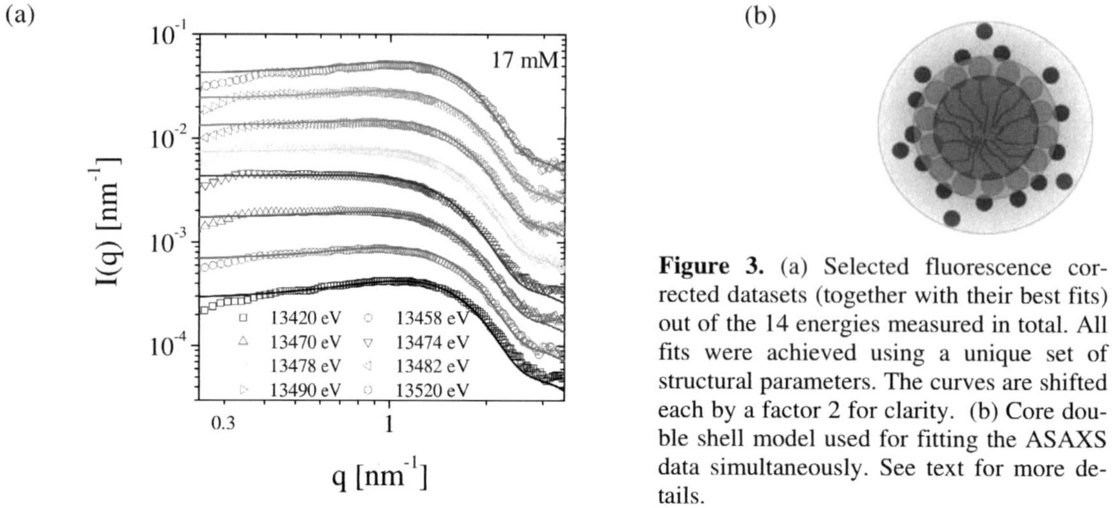

Figure 3. (a) Selected fluorescence corrected datasets (together with their best fits) out of the 14 energies measured in total. All fits were achieved using a unique set of structural parameters. The curves are shifted each by a factor 2 for clarity. (b) Core double shell model used for fitting the ASAXS data simultaneously. See text for more details.

The most interesting fit parameter is the spatial distribution of the bromide ions $v(r)$ (black dots) within the outer shell (grey). For data fitting a unique set of structural parameters [R, t, $v(r)$] was used and only the scattering contrast of the counterion layer [Eq. (5)] was varied according to the effective values of $f'(E)$ and $f''(E)$.

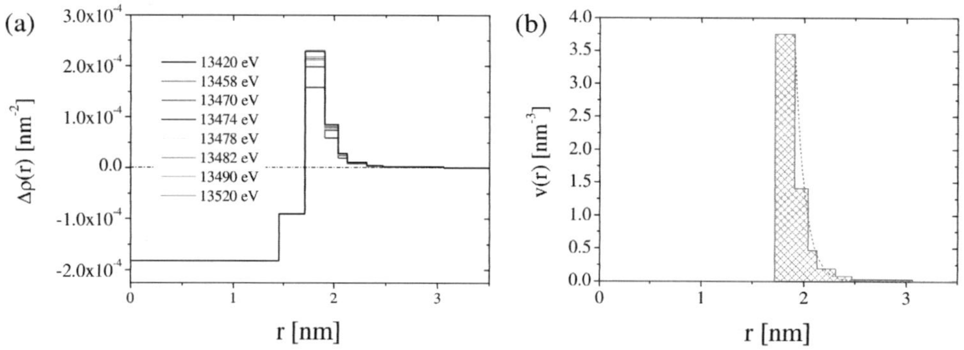

Figure 4. (a) Radial scattering length density profile to the best fits presented in Fig. 3. Only the scattering contrast within the counterion layer changes according to the energy dependence of $f'(E)$ and $f''(E)$. (b) Radial distribution of the counterions indicating that they are concentrated in a layer of ~0.34nm thickness with an exponential decay of the histogram to zero at a distance of 3.1nm from the center of the core.

The radial scattering length density profile $\Delta\rho(r)$ to the best fits presented before is shown in Fig. 4(a). The spatial distribution function of the Br counterions is presented in Fig. 4(b). The simultaneous fits of ASAXS data recorded below and above the Br K-edge confirm the derived fit results in [10]. The core radius is about R = 1.55nm and the thickness of the shell representing the head groups is t = 0.27nm. The distribution of the counterions is concentrated in a layer of ~0.34nm thickness (similar to the Stern layer) and decays exponentially to zero at a distance of 3.1nm from the center of the core, which is in agreement with the strong coupling of the counterions. The maximum number density of the ions is determined to be 3.8nm^{-3}, which corresponds to about 1/10 of the closest possible packing of the ions. The distribution function presented in Fig. 4(b) corresponds to about 72% of the bromide ions bound to the surfactant head groups. This result is consistent with the previous estimates of counterion association from neutron scattering [12] and electrochemical methods [13].

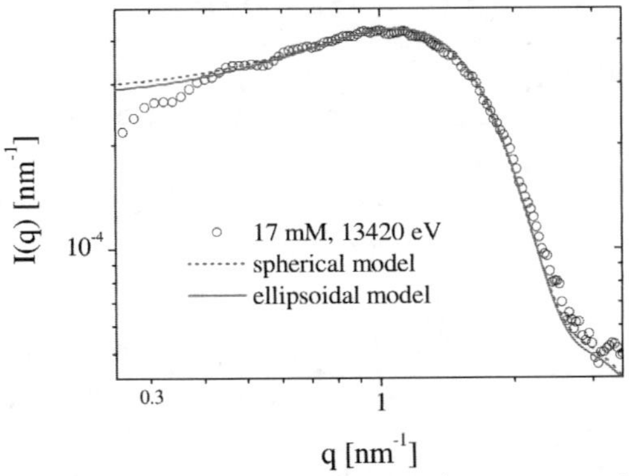

Figure 5. Comparison of scattering curves calculated from a spherical and ellipsoidal form factor model. The fitted polydispersity is 45% and < 20% for the two models, respectively.

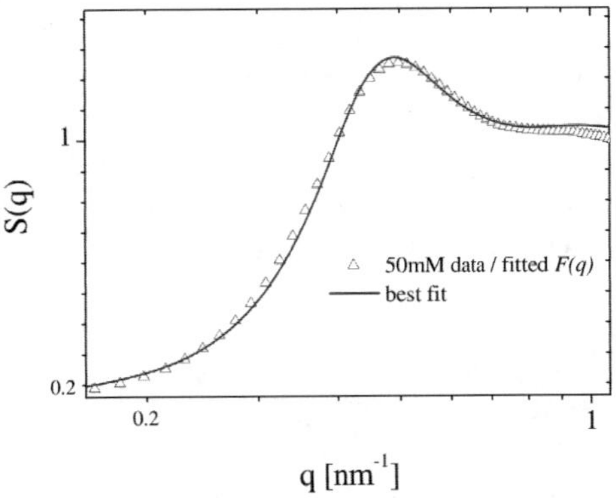

Figure 6. Averaged structure factor derived from measuring the sample with 50mM concentration at various energies and dividing by the corresponding fitted form factor. The fit of an effective structure factor for polydisperse homogeneous spheres shows good agreement.

Regarding the fit model, it should be considered that micelles have strong shape fluctuations resulting in an elongated form. Hence, they should be ideally fitted by an ellipsoidal form factor, which, however, necessitates a double numerical integration to take into account for size distribution and orientation of the particles. Therefore, a spherical model was chosen for the simultaneous fitting routine. Fig. 5(a) illustrates exemplarily that both a spherical form factor model and an ellipsoidal model can fit the experimental data recorded at 13420 eV. In the latter case, the orientational averaging of the ellipsoids also results in a smearing of the form factor oscillations and a smaller polydispersity is sufficient to describe the data. Therefore the polydispersity of the investigated system is less than 20% as compared to 45%, which is overestimated by the spherical model. The axes in the ellipsoidal core have a dimension of 1.34nm and 1.99nm.

The sample with a concentration of 50mM [Fig. 1(c)] shows a pronounced structure factor. The interaction between the particles was analysed dividing the data measured at various energies by the corresponding simulated form factor as derived from fitting the 17mM data [Fig. 3(a)]. The averaged structure term was fitted using an effective structure factor for polydisperse hard spheres [14]. The derived fit parameters give an effective radius of interaction of 3nm and a polydispersity of 19% which is in good agreement with the form factor fits.

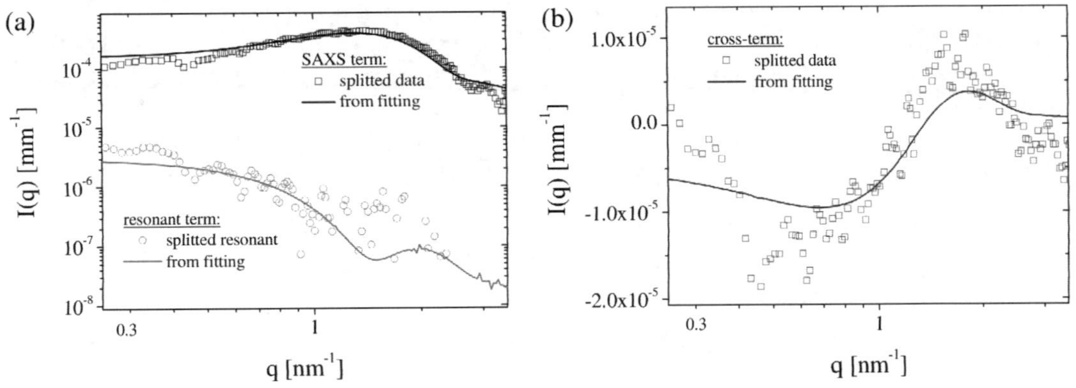

Figure 7. Comparison of the three partial intensities of SAXS term, cross-term and resonant term as derived from splitting the ASAXS data (17mM) presented in Fig. 1(b) (open symbols) and the best fit results shown in Fig. 3(a) (lines).

The derived best fit model was also compared to the three partial intensities which include the energy independent normal *SAXS term*, the *cross-term* involving the amplitudes of normal SAXS and the resonant scattering of the counterions, and the weak *resonant term* due to counterions. The splitted intensities were derived from the data recorded from the sample with 17mM concentration [Fig. 1(b)] according to Eq. (6) and numerically solving the corresponding system of equations. A routine implemented in the *SAXSutilities* package [15] was used for this purpose. Figure 7 shows the corresponding results. Open symbols represent data derived from ASAXS measurements whereas the continuous lines originate from the best-fit model presented in Fig. 3(a). Although the curves were not directly fitted to the partial intensities, they show very good agreement.

4. Discussion

The results presented in the previous section illustrate the power of ASAXS to probe the spatial distribution of counterions when the corresponding atomic absorption edge is experimentally accessible to quantitative SAXS measurements. The simultaneous analysis of the data recorded below and above the absorption edge provides additional constraints in data modelling. The fluorescence contribution close to and above the absorption edge can be corrected accurately for particulate systems. The spatial distribution of anomalous atoms derived by the direct modelling of the scattering curves and that deduced from the decomposed partial intensities is in good agreement. Both results reveal strong condensation of counterions on to the micelles. The form factor of the micelles can be described by a core-multi shell sphere model with an overestimated polydispersity or an ellipsoidal form with more realistic polydispersity. The structure factor can be modelled by a polydisperse effective hard-sphere system. The anomalous effect in the structure factor of interactions is weak compared to that in the form factor.

The European Synchrotron Radiation Facility is acknowledged for the provision of beam time.

References

[1] Holm C, Kekicheff P and Podgornik R 2001 *Electrostatic Effects in Soft Matter and Biophysics* (Kluwer Academic Publishers)
[2] Ballauff M and Jusufi A 2006 *Coll. Pol. Sci.* **284** 1303
[3] Stuhrmann HB 1985 *Adv. Polym. Sci.* **67** 123
[4] Naudon A 1995 *Modern Aspects of Small-Angle Scattering* (Kluwer Academic Publishers) 203
[5] Patel M, Rosenfeldt S, Ballauff M, Dingenouts N, Pontoni D and Narayanan T 2004 *Phys. Chem. Chem. Phys.* **6** 2962
[6] Das R, Mills TT, Kwok LW, Maskel GS, Millett IS, Doniach S, Finkelstein KD, Herschlag D and Pollack L 2003 *Phys. Rev. Lett.* **90** 188103
[7] Goerigk G, Schweins R, Huber K and Ballauff M 2004 *Europhys. Lett.* **66** 331
[8] Horkay F, Hecht AM, Rochas C, Basser PJ and Geissler E 2006 *J. Chem. Phys.* **125** 234904
[9] Dingenouts N, Patel M, Rosenfeldt S, Pontoni D, Narayanan T and Ballauff M 2004 *Macromolecules* **37** 8152
[10] Sztucki M, Di Cola E and T Narayanan 2010 *J. Appl. Cryst.* **43** 1479 and references therein
[11] Brennan S and Cowan PL 1992 *Rev. Sci. Instrum.* **63** 850
[12] Gorski N, Gradzielski M and Hoffmann H 1994 *Langmuir* **10** 2594
[13] Moulik SP, Haque ME, Jana PK and Das AR 1996 *J. Phys. Chem.* **100** 701
[14] Narayanan T 2009 *Lecture Notes in Physics, (Non-crystalline diffraction)* Ed. T. Ezquerra et al., 776 (Springer, Berlin-Heidelberg) pp 133-156
[15] Sztucki M and Narayanan T 2007 *J. Appl. Cryst.* **40** s459; SAXSutilities download page, http://www.sztucki.de/SAXSutilities

Journal of Physics: Conference Series **272** (2011) 012005 doi:10.1088/1742-6596/272/1/012005

Towards the Structural Characterization of Intrinsically Disordered Proteins by SAXS and MD Simulation

Tomotaka Oroguchi[1], Mitsunori Ikeguchi[1], and Mamoru Sato[1*]
[1]Graduate School of Nanobioscience, Yokohama City University, 1-7-29, Suehiro-cho, Tsurumi-ku, Yokohama 230-0045, Japan

E-mail: msato@tsurumi.yokohama-cu.ac.jp

Abstract. Dynamical structures of intrinsically disordered proteins (IDPs) and multi-domain proteins that include large ID regions between the domains are unable to be determined by such conventional methods as X-ray crystallography and electron microscopy. Small-angle X-ray scattering (SAXS) is suitable to determine low-resolution structures of proteins and protein complexes in solution, but the structural data on protein dynamics are averaged over the structural ensemble in protein solution. To overcome this problem, we have developed a novel method, named MD-SAXS, of the combined use of SAXS and molecular dynamics (MD) simulation to analyze protein dynamics in solution of multi-subunit protein complexes and multi-domain proteins toward the structural characterization of IDPs. Here we show validity of the method through the structural characterization of restriction Endonuclease EcoO109I.

1. Introduction

IDPs are characterized by lack of stable tertiary structure when the protein exists as an isolated polypeptide chain under physiological conditions. They adopt well-defined structures upon binding to their targets (coupled folding and binding). This is a novel target recognition mechanism unlike the "rock and key" and induced-fit models. IDPs are mostly found in eukaryotes and generally interact with several target molecules through the coupled folding and binding, thus playing important roles as protein hubs in intracellular networks [1]. However, structural characterization of IDPs by such conventional methods as X-ray crystallography and electron microscopy is difficult due to the extremely high flexibility of IDPs in solution, thus expecting development of a novel method to determine the IDP structure in solution at an atomic resolution. In view of this consideration, we have developed a novel method, named MD-SAXS of the combined use of SAXS and molecular dynamics (MD) simulation to analyze protein dynamics in solution of multi-subunit protein complexes and multi-domain proteins.

Restriction endonucleases (REases) provide anti-viral protection for bacteria by degrading the foreign DNA of invading bacteriophages [2]. These enzymes recognize specific nucleotide sequences and cleave both strands of DNA. EcoO109I is a type II REase isolated from *Escherichia coli* and recognizes 7 base pairs of double-stranded DNA (ds-DNA). The crystal structures of both DNA-free and DNA-bound forms of EcoO109I have been determined (Figure 1), and SAXS analysis of the DNA-free form have revealed that EcoO109I consists of dimerization and catalytic domains (Figure 1 D) and the homodimer (subunits A and B in Figure 1A) is the functional unit of the enzyme in solution

[2]. A comparison between the crystal structures of the DNA-free and DNA-bound forms of EcoO109I shows that the enzyme undergoes large structural changes upon DNA binding. These structural changes can be explained mainly by two movements. The first is the movement of the two catalytic domains in subunits A and B closer together, allowing interaction with DNA (Figure 1A-C). The second is the twisting motion between the two subunits in the counter-clockwise direction, making the two catalytic domains fit together on the major groove of DNA from both sides (Figure 1D-F).

In this study, we carried out a 150-ns MD simulation for the DNA-free form of the enzyme with the explicit water molecules in order to elucidate the relationship between the intrinsic dynamics of EcoO109I in solution and its structural changes occurring upon DNA binding [3]. From the trajectory of the MD simulation, the SAXS profile, taking scattering from explicit hydration water molecules into account, was calculated and compared with the experimentally observed SAXS profile. The time-resolved analysis of the radius of gyration estimated from the Guinier plot demonstrates that the structure of EcoO109I is considerably flexible. Analyses of atomic fluctuations within the subunit and in the dimer identified two rigid cores that undergo rigid-body motions relative to each other. The motions of the two rigid cores were characterized using principal component analysis, and were compared with the DNA-induced structural changes observed in the crystallographic study [2]. Motions in the catalytic domains were also examined using principal component analysis.

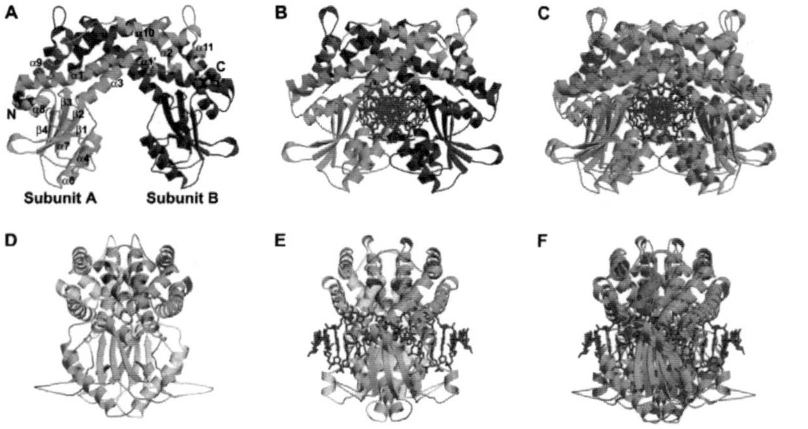

Figure 1. The crystal structures of (A) DNA-free and (B) DNA-bound forms colored according to subunits A (green) and B (purple). Panels (D) and (E) show a side view of the DNA-free and DNA-bound forms, respectively. Panels (C) and (F) show front and side views of the superimposition of these two forms, respectively.

2. Methods

2-1: MD simulations

All simulations were performed at 298 K using MARBLE [6]. The initial structure for the simulation of EcoO109I was generated from the crystal structure (protein data bank (PDB) accession code: 1WTD [2]). The protein and crystal water molecules were immersed in a water box, followed by energy minimization of 5000 steps of the whole system. The total number of atoms was 108,249. Two series of product runs were performed under the NPT condition with $T = 298.15$ K and $P = 1.0$ atm: (i) a 150-ns run without any restraints; and (ii) a 20-ns run with restraints on the positions of all the protein atoms to the crystal coordinates. The more details about MD procedures are described in the literature [3].

2-2: Calculation of the SAXS profile

The experimental data of the SAXS profile were taken from the previous study [2]. The calculation procedure is as follows: (i) MD simulations of the pure solvent system V are performed in addition to simulations of the protein solution (protein molecules in solution) system U and (ii) the MD-SAXS excess intensity I is obtained by a subtraction of the pure solvent scattering intensity I^V from the protein solution intensity I^U. This procedure is similar to that used in typical SAXS experiments: the two sets of data I^V and I^U are collected and the difference between them is used for further data analysis. I^V and I^U are calculated by using only atoms included in region "v", which corresponds to the area inside a sphere centered at the center of mass of the protein molecule (Figure 2A). As proven in the literature [3], the scattering from region "bo" corresponding to the area outside the sphere and the cross term of scattering between region v and region bo are equivalent between the protein solution and the pure solvent when the radius R of region o is large enough to include not only the entire solvation layer around the protein but also some of the bulk region. Therefore, those scattering terms become zero by subtraction. Then, the objective MD-SAXS excess intensity $I(Q)$ is expressed as

$$I(Q) = I^U(Q) - I^V(Q) \tag{1}$$

where

$$I^\eta(Q) = \left\langle \tilde{I}(Q) \right\rangle_\eta \tag{2}$$

$(\eta = U, V)$.

$\mathbf{Q}$ is the scattering vector, which is related to the scattering angle 2θ and the wavelength of the incident radiation λ by the equation $Q = |\mathbf{Q}| = 4\pi \sin\theta/\lambda$. $\langle\ \rangle_U$ and $\langle\ \rangle_V$ represent the configurational averages over the ensembles of the protein solution and the pure solvent, respectively. $\tilde{I}(Q)$ in these brackets is the instantaneous scattering intensity, which is calculated for each snapshot in the trajectories of the protein solution and pure solvent systems using by the following equation:

$$\tilde{I}(Q) = \left\langle \int_{\mathrm{v}} d^3\mathbf{r}' \int_{\mathrm{v}} d^3\mathbf{r} (\tilde{\rho}(\mathbf{r}') - \rho_0)(\tilde{\rho}(\mathbf{r}) - \rho_0) e^{-i\mathbf{Q}\cdot(\mathbf{r}-\mathbf{r}')} \right\rangle_{\Omega_\mathbf{Q}} \tag{3}$$

where $\tilde{\rho}(\mathbf{r})$ is the electron density at position $\mathbf{r}$, and ρ_0 is the average electron density of the bulk in the protein solution. In this study, the time-averaged density of the solvent in the region bo was used as ρ_0. $\langle\ \rangle_{\Omega_\mathbf{Q}}$ denotes the orientational average. For the fast computation, equation (3) is calculated using the multipole expansion:

$$\exp(-i\mathbf{Q}\cdot\mathbf{r}) = 4\pi \sum_{l=0}^{\infty} \sum_{m=-l}^{l} i^l j_l(Qr) Y_{lm}^*(\Omega_\mathbf{r}) Y_{lm}(\Omega_\mathbf{Q}), \tag{4}$$

where j_l and Y_{lm} denotes a spherical Bessel function and a spherical harmonic function, respectively. Y_{lm}^* is a complex conjugate of Y_{lm}. $\Omega_\mathbf{r}$ is the solid angle in real space, $\mathbf{r} = (r, \Omega_\mathbf{r})$. Then, equation (3) is rewritten as:

$$\begin{aligned}\tilde{I}(Q) = {} & 4\pi \sum_{l=0}^{\infty} \sum_{m=-l}^{l} \left| i^l \sum_{k}^{N} f_k(Q) j_l(Qr_k) Y_{lm}^*(\Omega_{\mathbf{r}_k}) \right|^2 - 2\rho_0 \sum_{k}^{N} f_k(Q) \int_{\mathrm{S}} dS(\mathbf{s}_k \cdot \mathbf{n}) \frac{j_1(Qs_k)}{Qs_k} \\ & + \rho_0^2 \int_{\mathrm{S}} dS' \int_{\mathrm{S}} dS \left[\frac{(\mathbf{n}'\cdot\mathbf{n})}{Q^2} \cdot \frac{j_1(Qs')}{Qs'} - \{(\mathbf{s}'\cdot\mathbf{n}')(\mathbf{s}'\cdot\mathbf{n})\} \frac{j_2(Qs')}{(Qs')^2} \right].\end{aligned} \tag{5}$$

The first term of the right hand side of equation (5) denotes the scattering from all explicit atoms within the region v, the second term denotes the cross term of the scattering between these atoms and the bulk solvent in the region bo, and the third term originates from the scattering from the infinite bulk. N is the number of atoms within the region v. $\mathbf{r}_k$, and $\Omega_{\mathbf{r}_k}$ is the position and the solid angle in real space of the k-th atom, respectively, and $\mathbf{s}_k$ and $\mathbf{s}'$ denotes $\mathbf{r} - \mathbf{r}_k$ and $\mathbf{r} - \mathbf{r}'$, respectively. $f_k(Q)$ denotes the atomic form factor of the k-th atom. S represents the closed surface of the region v, and $\mathbf{n}$ is the unit normal vector outward from the infinitesimal surface element dS of the surface S. The more detailed derivations and definitions of the equations (3) and (5) are described in the literatures [3-5].

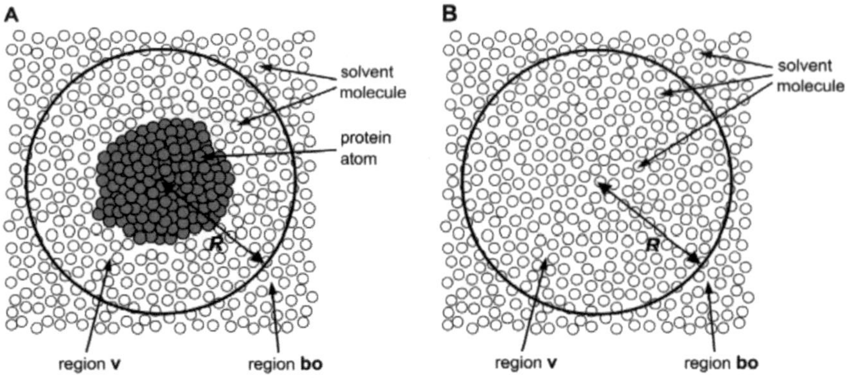

Figure 2. Illustration of dividing the total volume of a system into two regions 'v' and 'bo', showing a dilute solution of a protein (A) and pure solvent (B).

3. Results

During the 150-ns MD simulation of the DNA-free form of EcoO109I, the whole structure fluctuated largely. The root-mean-square displacement of the C^{α} atoms (C^{α}-RMSD) from the crystal structure of the DNA-free form reached ~5.0 Å within the first 20 ns of the simulation. However, the C^{α}-RMSD of individual rigid cores, i.e., the elements of rigid-body motions, was ~1.0 Å during the simulation, indicating that the rigid cores were stable during the simulation, and that there were certain rigid-body motions between them.

3-1: Comparison of the MD Simulations with the Experiment in SAXS data

To test the validity of the structural ensemble sampled by the MD simulation, we calculated the SAXS profile from all the snapshots in the trajectory, taking scattering from the explicit hydration water molecules into account and compared it with the experimental profile. As shown in Figure 3A, the simulation-derived profile was in agreement with the experimentally observed SAXS profile.

In the range of $R_g Q < 1.3$, the scattering intensity $I(Q)$ is approximated to (Guinier approximation),

$$I(Q) \approx I(0)\exp(-Q^2 R_g^2/3), \qquad (6)$$

where $I(0)$ and R_g are the scattering intensity at $Q = 0$ (zero-angle) and the radius of gyration, respectively, which are estimated from $\ln[I(Q)]$ versus Q^2 plot (Guinier plot) [7]. Figure 3B compares the Guinier plots of the simulation-derived and experimentally observed SAXS profiles, showing that the R_g (referred to as 'protein-water R_g' or simply 'R_g') estimated from the SAXS profile calculated from the structural ensemble sampled by the MD simulation (R_g = 28.2 Å) was in agreement with that obtained from the experimentally observed SAXS profile (R_g = 28.1 ± 0.3 Å).

We then calculated time-evolution of the protein-water R_g to investigate the fluctuation of the overall structure in the 150-ns simulation. R_g (the cyan curve in Figure 3C) exhibited large fluctuation with a period of about 50-100 ns. There was an approximately 1.8-Å difference in R_g between the most expanded and closed conformations; however, R_g averaged over all time-windows (the blue line in Figure 3C) is in agreement with R_g obtained from the experimental profile (the pink horizontal line).

We also calculated time evolution of another R_g (referred to as 'protein R_g'), which was obtained only from the protein structure (the black curve in Figure 3C). The time course of the protein R_g clearly shows almost the same time dependency as that of the protein-water R_g; this indicates that the fluctuation of the protein-water R_g reflects the fluctuation of the protein moiety, and a 2-Å difference between the protein-water R_g and the protein R_g is attributable to the hydration shell on the surface of the protein molecule.

In order to analyze the effects of structural changes on the protein-water R_g during the 150-ns MD simulation, we performed another 20-ns MD simulation in which positions of the protein atoms were restrained to the original crystal positions and only the solvent atoms were sampled (referred to as the 'restraint-MD simulation'). As shown in Figure 3C (the yellow-green curve), we calculated the protein-water R_g for the time window of 500 ps shifting from t = 0-20 ns for the restraint-MD simulation. In this simulation, the protein-water R_g was ~27.6 Å, and the average of the protein-water R_g (the green horizontal line in Figure 3C) was still smaller than the experimental value. The difference in the protein-water R_g between the restraint-MD simulation and the experimental value indicates that EcoO109I adopts a more expanded conformation in solution, as observed in the 150-ns MD simulation, than in crystalline state. All these results show the validity of the structural ensemble sampled by the 150-ns MD simulation. In the following section, we investigate the detailed dynamics of EcoO109I during the 150-ns MD simulation.

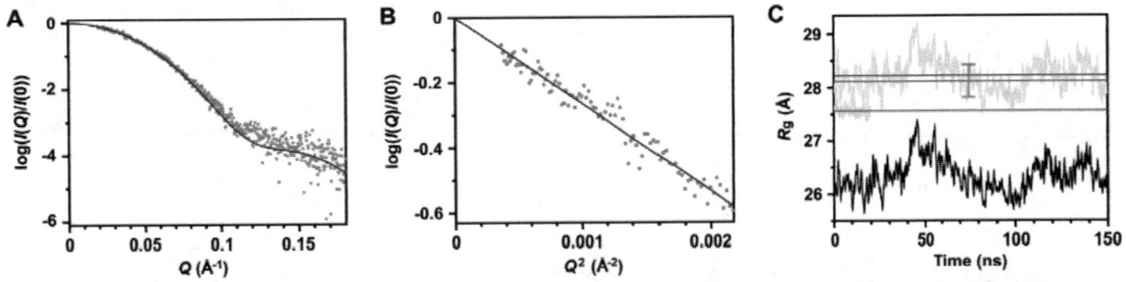

Figure 3. Comparison of the experimental and simulation-derived SAXS profiles of the DNA-free form of EcoO109I in solution. The SAXS profiles from the experiments (pink) and the 150-ns MD simulation (blue) are shown in Logarithmic plot (A) and Guinier plot (B). Panel (C) shows time evolution of the protein-water R_g for the 150-ns MD trajectory (cyan) and the 20-ns restraint-MD trajectory (yellow-green). The pink line shows the protein-water Rg estimated from the experiment. The protein R_g without water molecules was also calculated for the 150-ns trajectory (black).

3-2: Intrinsic Dynamics of DNA-free Form of EcoO109I

For further analyses of structural fluctuations of EcoO109I in the simulation, the principal component analysis was applied for the 150-ns trajectory. The first two principal modes accounted for the dominant portion of the total fluctuations of this protein (Figure 4A).

The eigenvector of the first principal mode represents the open-close motion (the red arrows in Fig. 4 *D*). The time course of projections onto the first principal mode is shown in Figure 4B, in which positive and negative projection values correspond to open and closed conformations, respectively. Because the R_g of the open conformations were larger than those of the closed conformations, there was a significant correlation in the time courses between the projections onto the first principal mode and the time-dependent R_g shown in Figure 3C (the correlation coefficient = 0.74). Moreover, we found that the open-close motion represented by the first principal mode changed the width of the intervening space between the two catalytic domains, d_{bc} (Figure 4B-C). In Figure 6E, the superposition of the structures with the largest and smallest projections onto the first principal mode is shown. The structure with the largest projection value (green in Figure 4E) has a d d_{bc} of 20.9 Å, which is slightly larger than the diameter of DNA (20 Å). In contrast, the structure with the smallest projection value (pink in Figure 4E) has a d_{bc} of 5.2 Å, which is close to the crystal structure in the DNA-bound form (2.7 Å). Because the d_{bc} of the crystal structure in the DNA-free form is 6.4 Å (smaller than t DNA), the open-close motion represented by the first principal mode is necessary for the enzyme to access DNA.

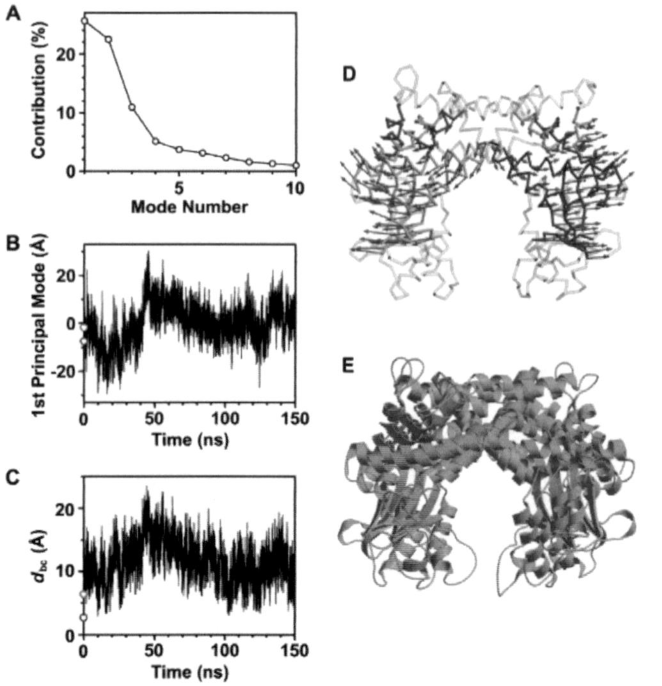

Figure 4. Principal component analysis for motions in the 150-ns MD trajectory. Panel (A) shows individual contributions of the first 10 principal components to the total fluctuations. (B) Snapshots in the trajectory were projected onto the first principal mode. The DNA-free (red open circle) and DNA-bound (blue open circle) crystal structures are also projected onto the first principal mode. (C) The distance between the two catalytic domains, d_{bc}, in the 150-ns simulation is shown. (D) The amplitude and direction of the first principal mode are represented by red arrows on the average structure. (E) The superposition of the structures with the largest projection (green) and smallest projection (pink) onto the first principal mode.

The eigenvector of the second principal mode represents the twisting motion between the two subunits in the counter-clockwise direction (the red arrows in Figure 5B). The superposition of the structures with the largest and smallest projections is also shown in Figure 5C. As seen in the comparison of Figure 1F with Figure 5C, the direction of the second principal mode was significantly similar to the structural changes occurring upon DNA binding. Indeed, the correlation coefficient in the directions between the second eigenvector and the structural change in the DNA-free and DNA-bound crystal structures is 0.90 (Figure 5B). This indicates that structural changes upon DNA binding clearly correlate with the intrinsic dynamics of EcoO109I. In other words, the relative arrangement between the two subunits is specifically flexible in the counter-clockwise direction in the DNA-free state, and this flexibility effectively facilitates structural changes upon DNA-binding.

4. Conclusion

To elucidate the relationship between the intrinsic dynamics of EcoO109I and its function, we performed a 150-ns MD simulation of the DNA-free form of the enzyme and examined the resultant trajectory in detail. We calculated the SAXS profile from all the snapshots in the trajectory, taking scattering from the explicit hydration water molecules into account. The calculated SAXS profile is in agreement with the experimentally observed SAXS profile, and a difference between the protein-water R_g and the protein R_g is 2 Å and coincides with the hydration shell width on the surface of the protein molecule. These results support the validity of the structural ensemble sampled by the present simulation. In addition, a large fluctuation in R_g was observed in the simulation, which indicates considerable flexibility in the overall structure of the enzyme.

From the principal component analysis applied to the trajectory, we have identified intrinsically functional motions in the DNA-free form of EcoO109I. The motion represented by the first principal mode is necessary for the intervening space between the catalytic domains to be accessed by DNA. The motion represented by the second principal mode has a significant correlation with the structural changes upon DNA binding, and is therefore necessary for EcoO109I to effectively achieve functionality. These two modes are dominant fluctuations in the DNA-free state.

Our present study using the MD-SAXS has revealed that the combination of SAXS and MD simulations is very powerful to identify the structural ensemble at an atomic resolution, and thus elucidate the functional motions of the enzyme. In recent studies, we have extended our MD-SAXS approach to make it possible to apply it to the flexible proteins, the crystal structures of which are very difficult to be obtained. One example of such a protein is the e-subunit of F1-ATPase, which takes the extended form in the absence of ATP. We have revealed the structural ensemble of the ATP-free form of the e-subunit by our extended MD-SAXS approach (data not shown). Our approach can be also applied to determine multi-subunit protein complexes and multi-domain proteins in solution at an atomic resolution. We are now extending our method to analyze the extremely flexible proteins, IDP.

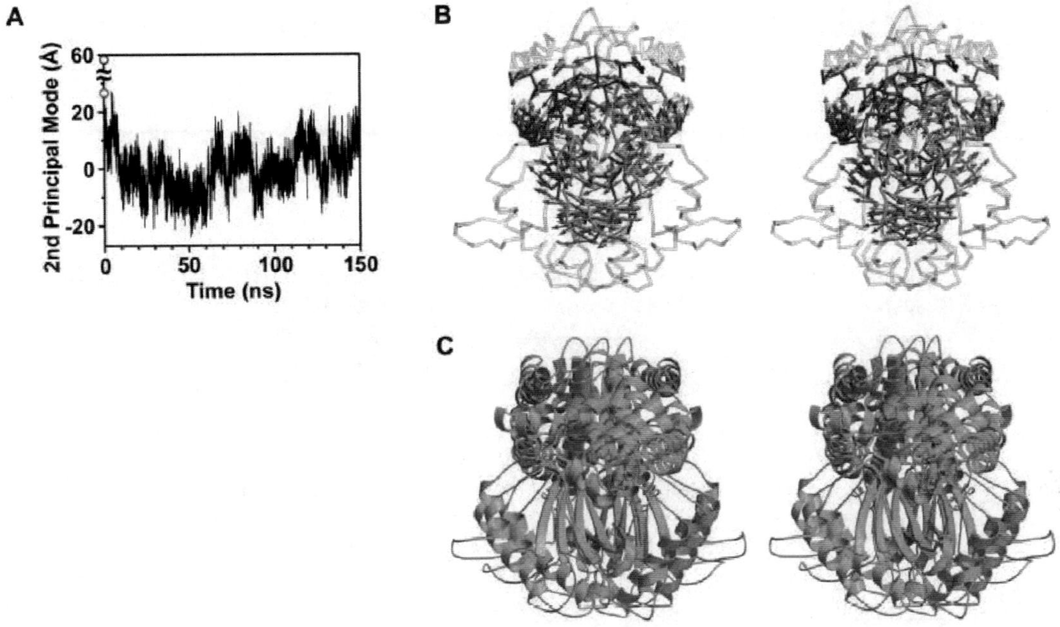

Figure 5. The second principal mode. (A) Snapshots in the trajectory were projected onto the second principal mode. (B) The amplitude and direction of the second principal mode are represented by red arrows on the average structure. The structural change between the DNA-free and bound forms is represented by cyan arrows. (C) The superposition of the structures with the largest and smallest projections onto the second principal mode.

Acknowledgements
This study was supported in part by Grants-in-Aid for Scientific Research from the Ministry of Education, Culture, Sports, Science and Technology (MEXT), grants from the Japan Science and Technology Agency, and Research and Development of the Next-Generation Integrated Simulation of Living Matter, a part of the Development and Use of the Next-Generation Supercomputer Project of MEXT. All computations were carried out at Tsurumi campus of the International Graduate School of Arts and Sciences, Yokohama City University, Yokohama, Japan.

References
[1] Uversky V N and Dunker A K 2010 Biochimica et Biophysica Acta **1804** 1231
[2] Hashimoto H, Shimizu T, Imasaki T, Kato M, Shichijo N, Kita K and Sato M 2005 *J. Biol. Chem.* **280** 5605
[3] Oroguchi T, Hashimoto H, Shimizu T, Sato M and Ikeguchi M 2009 *Biophys. J.* **96** 2808
[4] Seki Y, Tomizawa T, Khechinashvili N N and Soda K 2002 *Biophys. Chem.* **95** 235
[5] Soda K, Miki Y, Nishizawa T and Seki Y 1997 *Biophys. Chem.* **65** 45
[6] Ikeguchi M 2004. *J. Comput. Chem.* **25** 529
[7] Guinier A 1939 *Ann. Phys.* **12** 161

Journal of Physics: Conference Series **272** (2011) 012006 doi:10.1088/1742-6596/272/1/012006

Ultra small angle X-ray scattering studies on density heterogeneity of linear low density polyethylene

M Takenaka[1,2], S Fujii[1], S Nishitsuji[1], N Yagi[3], Y Suzuki[3], and A Takeuchi[3]

[1]Department of Polymer chemistry, Graduate School of Engineering, Kyoto University, Kyotodaigakukatura, Nishikyo-ku, Kyoto, 615-8510, JAPAN

[2]Structural Materials Science Laboratory, SPring-8 Center, RIKEN Harima Institute Research, 1-1-1, Kouto, Sayo-cho, Sayo-gun, Hyogo 679-5148, JAPAN

[3]SPring-8 JASRI, □1-1-1, Kouto, Sayo-cho, Sayo-gun, Hyogo 679-5198, JAPAN

E-mail: takenaka@alloy.polym.kyoto-u.ac.jp

Abstract. We have investigated density heterogeneity of linear low density polyethylene (LLDPE) in the order of submicron scale with 2 dimensional ultra small angle X-ray scattering (2D USAXS). We found that power-law behaviour in small q- region of isotropic scattering patterns of LLDPE studied here, indicating that the density heterogeneity obeys mass fractal. Under stretching, isotropic 2D USAXS patterns of some LLDPE samples were transformed into butterfly patterns caused by heterogeneous deformation.

1. Introduction

Polyethylene (PE) is one of the common polymers used widely in our daily life. PE is a crystalline polymer and its physical properties strongly depends on the morphologies of crystalline or heterogeneity of density. PE forms a spherulite consisting of a hierarchical structure[1,2]. Among each level of structures, crystalline unit cell structure and lamellar structure have been well studied by using wide angle X-ray scattering (WAXS), and small angle X-ray scattering (SAXS) and obtained the characteristic parameters such as the degree of crystallinity, domain spacing of lamellar structures[3]. On the other hand, though the density heterogeneity in the order of sub micron scale in spherulite is thought to affect the mechanical properties of crystalline polymers, density heterogeneity in spherulite such as fibril structure has not been well investigated yet. This is because light scattering (LS) method, which is commonly used to observe the length scale from 100 nm to 1μm, can not be applied to investigate the density fluctuations of crystalline polymer having optical anisotropy in its crystal. However, the ultra small angle X-ray scattering (USAXS) camera with long path length and the synchrotron X-ray source has become available recently and it enables us to measure 2-dimensional (2D) USAXS pattern so that we can investigate anisotropic structure in submicron scale. In this study, we, thus, aim at the exploration of density heterogeneity of linear low density polyethylene (LLDPE) samples in the order of sub micron scale. Moreover, we would like to show how the stretch affect the density heterogeneity in the LLDPE samples. In our previous study[3] on the density heterogeneity in the order of submicron scale for high density polyethylene, we found the isotropic patterns was transfored into butterfly pattern by stretching. The spatial heterogeneity of stress field or viscoelastic effects[4] was found to induce the butterfly pattern, similar to the case of the butterfly pattern found in the chemical gel. We investigate whether the butterfly pattern appears in the case of LLDPE.

2. Experimental Section

2.1. Sample

We used three kinds of LLDPE : LLD-a(density 0.918g/ml(25°C), Melt Flow Index (MFI) 1.0g/10min (190°C /2.16kg), ALDRICH 428078), LLD-b(density 0.926g/ml(25°C), MFI 50.0g/10min (190°C /2.16kg), NEO-ZEX 25500J), LLD-c(density 0.931g/ml(25°C), MFI 26.0g/10min (190°C /2.16kg), NEO-ZEX 30200J). LLD-a includes hexene as a comonomer while butene is contained in LLD-b and LLD-c as a comonomer. The specimens are prepared by melt press method: First, the pellets of the samples were melted in a hot-press machine for an hour at 150°C. Subsequently the samples were crystallized for 2 hours at 120°C for LLD-a and 110°C for LLD-b and LLD-c. Finally, the specimens were quickly cooled to room temperature with iced water. The sample thickness was set to be 2mm. The specimens were subjected to uniaxial extension at room temperature with the stretch rate being about 2.0×10^{-3} sec^{-1}. The elongation ratio ε is defined by

$$\varepsilon = L / L_0, \qquad (1)$$

where L and L_0 are, respectively, the length of the sample after and before stretch.

2.2. USAXS experiment

We employed long path length (~150m) and the synchrotron X-ray source at BL20XU, SPring-8 to investigate the density fluctuations in the order of the submicron scale. By using the setup at BL20XU, we can observed 2D scattering pattern at very small wave number q-region; $0.005 < q < 0.02$ nm^{-1}, where q is defined by

$$q = (4\pi/\lambda)\sin(\theta/2) \qquad (2)$$

with λ and θ being wavelength of the incident beam and scattering angle, respectively. In this experiment, we used λ=0.05 nm. To detect the scattered 2D intensity quantitatively, we employed the CCD camera with image intensifier[5]. The detail of the instruments was discussed elsewhere[6]. We

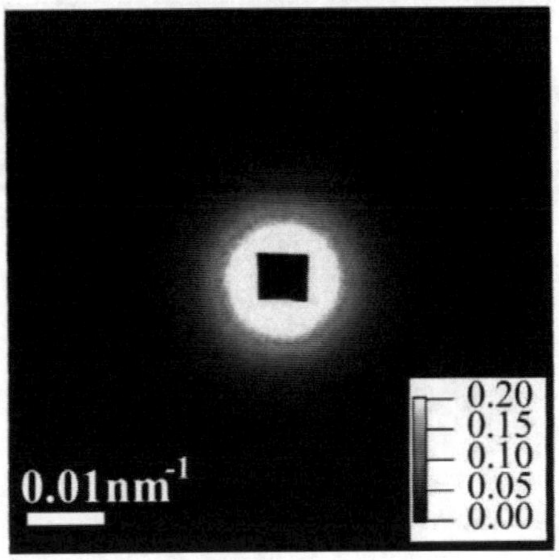

Figure 1 2D-USAXS patterns of LLD-a at ε=1.0

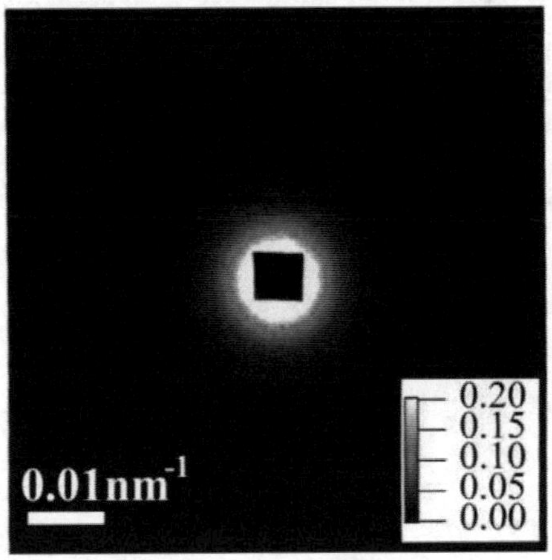

Figure 2 2D-USAXS patterns of LLD-b at ε=1.0

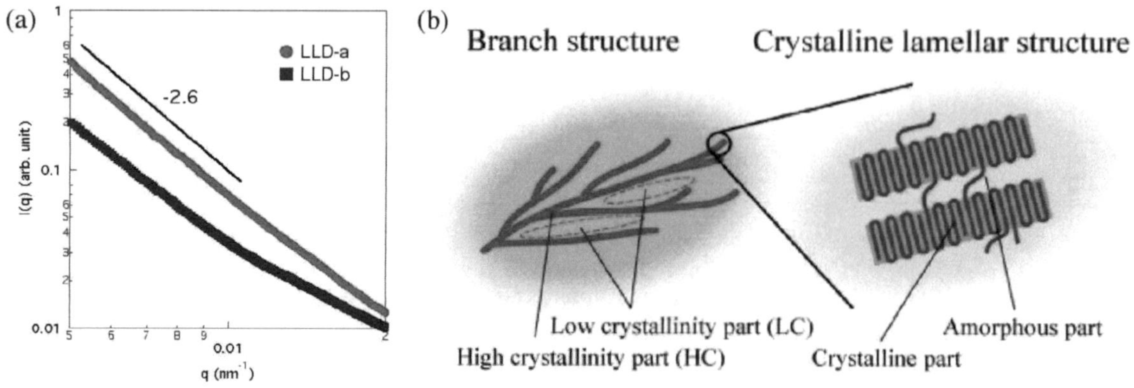

Figure 3 (a) the circularly-averaged scattering profile $I(q)$ of LLD-a and LLD-b without stretching plotted as a function of q. (b) The schematic illustration of the mass-fractal structure in the PE spherulite

conducted the 2D USAXS measurement at room temperature.

3. Results and discussion

3.1. 2D USAXS patterns without stretching

Figures 1 and 2 show the 2D USAXS patterns for LLD-a and LLD-b without stretching. The 2D USAXS shows isotropic scattering patterns. Similar to the case of LLD-a and LLD-b, the isotropic scattering patterns can be observed in LLD-c. These results indicate that the density heterogeneity in the order of submicron scale exists in LLDPE. It should be noted that the scattered intensity at such low q-region or sub-micron scale indicates the existence of larger scale density fluctuations than those of lamellar structures. In Fig.3(a), we show the circularly-averaged scattering profile $I(q)$ of LLD-a and LLD-b without stretching plotted as a function of q. We found the power-law behavior expressed by

$$I(q) \sim q^{-2.6}. \qquad (3)$$

The 2.6 value of the exponent is between 1 and 3, indicating that the large scale density fluctuations of spherulite obey mass-fractal as shown in Figure 3. The mechanism to form the observed mass-fractal structure is not clear. However, as shown in the studies by Vaughan and Basett[7] with electron micrograph, a sheaf-like structure formed by an aggregate of layers appears at the center of the spherulite in early stage of the growth of spherulites and then fibrils, composed of aggregates of layers, branch and splay to keep the increasing surface of the spherulite. Thus, one of the possible origin of the mass-fractal density fluctuations is the imperfection of volume filling by branching and splaying fibrils and the fibrils distribute in amorphous phase or less crystalline phase with mass-fractal feature. It also should be noted that the 2.6 value of the exponent is similar to that obtained by percolation theory[8]. In order to clarify the mechanism of the formation of the mass fractal structure, we need to investigate the time change in the scattered intensity during crystallization process with 2D USAXS.

3.2. 2D USAXS patterns after stretching

2D USAXS patterns for LLD-a and LLD-b after stretching are shown in Figures 4 and 5. In the case of LLD-a, the scattering patterns are elongated perpendicular to the stretch direction and becomes ellipsoidal patterns, indicating that the branch structure is elongated homogeneously. On the other hand, the scattering pattern of LLD-b exhibits butterfly patterns elongated parallel the stretch direction. This kind of butterfly pattern can be observed in the scattering patterns of shear-induced phase separation in semi-dilute polymer solutions[9].

Journal of Physics: Conference Series **272** (2011) 012006 doi:10.1088/1742-6596/272/1/012006

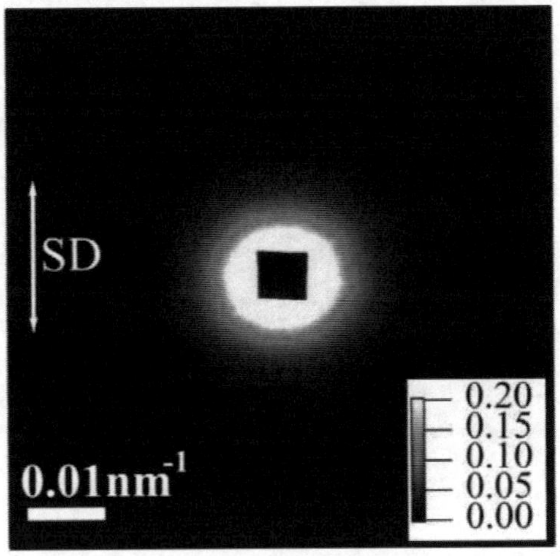

Figure 4 2D-USAXS patterns of LLD-a at ε=1.10. Arrow indicates the stretch direction.

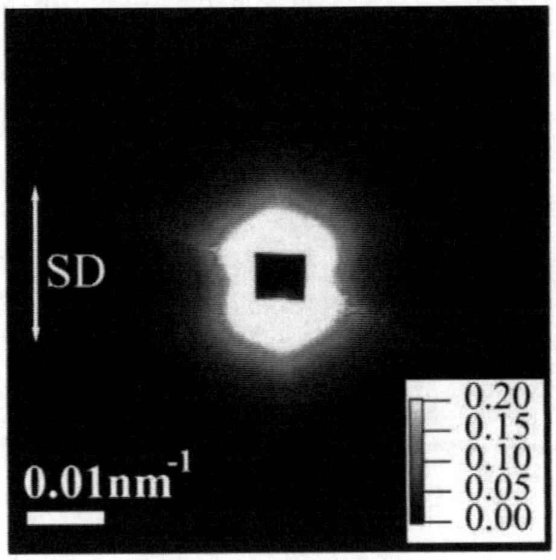

Figure 5 2D-USAXS patterns of LLD-b at ε=1.20. Arrow indicates the stretch direction.

In semi-dilute polymer solutions, the spatial heterogeneity of stress field induces associating with the concentration fluctuations since the dynamical properties of each component (polymer and solvent) are asymmetric. The heterogeneity of the stress field affects the free energy functional of the semi-dilute polymer solutions and causes the phase separation under shear field. Similarly, crystalline LLDPE has spatial heterogeneity of stress field associating with degree of crystallinity. Actually, the density fluctuations can be observed in 2D USAXS patterns of LLDPE without stretching. Under stretching, the heterogeneity of the stress field induces the density fluctuations and the butterfly patterns appear in the 2D USAXS.

There are two possible origins to explain the difference in behaviors in the scattered intensity under stretch between LLD-a and LLD-b and –c. One is the difference of molecular weight. In the case of LLD-a, MFI is much smaller than those of LLD-b and LLD-c, implying that the molecular weight of LLD-a is much larger than those of the others. Thus, the dynamical asymmetry between crystalline phase and amorphous phase of LLD-a is much smaller than those of the others and the small heterogeneity of the stress field did not induce butterfly pattern in LLD-a. Another possible origin is the difference of the comonomer. The comonomer of LLD-a is hexene while the comonomer of LLD-b and LLD-c is butene. The difference of comonomer may cause the difference of the mechanical proerties of crystalline or amorphous phases. We need to investigate other LLDPE samples with 2DUSAXS further.

4. Conclusion

We have investigated density heterogeneity of linear low density polyethylene (LLDPE) in the order of submicron scale with 2 dimensional ultra small angle X-ray scattering (2D USAXS). We found that power-law behavior in small q-region of isotropic scattering patterns of LLDPE studied here, indicating that the density heterogeneity obeys mass fractal. Under stretching, isotropic 2D USAXS patterns of some LLDPE samples were transformed into butterfly patterns caused by heterogeneous deformation associating with the heterogeneity of the stress field induced by the dynamical asymmetry between crystalline phase and amorphous phase. In the case of high molecular weight LLDPE, the stretch induces the homogeneous or affine deformation and the butterfly patterns can not be observed. This is because the dynamical asymmetry is not enough to cause the density fluctuations.

References

[1] Gerasimov V I, Ya Genin V, Ya Tsvankin D, 1974 *J. Polym. Sci.: Polym. Phys. Ed.*, **12**, 2035

Journal of Physics: Conference Series **272** (2011) 012006 doi:10.1088/1742-6596/272/1/012006

[2] Bowden P B, Young R J, *J. Mater. Sci.*, 1974 **9**, 2034
[3] Takenaka M, Shimizu H, and Nishitsuji S, 2007 *Physical Review E*, **75**, 061802
[4] Furukawa A and Tanaka H, 2006 Nature **443**, 434
[5] Amemiya Y, Ito K, Yagi N, Asano Y, Wakabayashi K, Ueki T, and Endo T, 1995 *Review of Scientific Instruments*, **66**, 2290
[6] Yagi N, Inoue K, 2003 *Journal of Applied Crystallography*, **36**, 783
[7] Strobl G, 1997 *The Physics of Polymers*, Springer, Berlin
[8] Martin JE, D. Adolf D, and Wilcoxon JP, 1989 *Phys. Rev. A*, **39**, 1325
[9] Hashimoto T, Fujioka K, 1991 *J. Phys. Soc. Jpn.*, **60**, 356

Journal of Physics: Conference Series **272** (2011) 012007 doi:10.1088/1742-6596/272/1/012007

Effects of blending poly(D,L-lactide) with poly(ethylene glycol) on the higher-oder crystalline structures of poly(ethylene glycol) as revealed by small-angle X-ray scattering

N D Tien[1], T P Hoa[2], G Kimura[1], Y Yamashiro[1], H Fujiwara[1], M Mochizuki[3], S Sasaki[1] and S Sakurai[1,3]

[1]Department of Biobased Materials Science, Kyoto Institute of Technology, Matsugasaki, Sakyo-ku, Kyoto 606-8585, Japan, Tel: 075-724-7890, Fax: 075-724-7800, E-mail: dungtien.hut@gmail.com, shin@kit.ac.jp
[2]Polymer Center, Hanoi University of Technology, Hanoi, Vietnam
[3]Center for Fiber and Textile Science, Kyoto Institute of Technology, Japan

Abstract. Effects of blending poly(lactic acid) (PLA) with poly(ethylene glycol) (PEG) on higher-order crystalline structures of PEG were examined using small-angle X-ray scattering (SAXS). For this purpose, the fact that two polymers are both crystalline makes situtation much complicated. To simplify, non-crystalline PLA is suitable. Thus, we used poly(D,L-lactic acid) (DLPLA), which is random copolymer comprising D- and L-lactic acid moieties. Multiple scattering peaks arising from the regular crystalline lamellar structure were observed for the PEG homopolymer and the blends. Surprisingly, the structure is much more regular for the blend DLPLA/PEG at composition of 20/80 wt.% than for the PEG homopolymer. Also for this blend sample as well as for a PEG homopolymer, very peculiar SAXS profiles were observed just 1^{0}C below T_m of PEG. This is found to be a particle scattering of plate-like objects, which has never been reported for polymer blends or crystalline polymers. Futhermore, it was found that there was strong hysteresis of the higher-order structure formation.

1. Introduction

Poly(lactic acid) (PLA) is at present the most promising and interesting biodegradable polymers in the family of aliphatic polyester. Polylactides have a chiral carbon in the lactic acid unit, giving rise to two stereoforms, poly-L-lactide (PLLA) and poly-D-lactide (PDLA). Both are highly crysatalline in contrast to racemic poly-DL-lactide (DLPLA) which is amorphous consisting of both D- and L-lactic acid [1].

Because of rather poor properties such as poor thermal stability[2,3] and highly sensitive to heat[3,4], poor long-term durability[2,5] and lower impact strength[5,6] several approaches can be used to improve the properties of PLA[3,7]. Blending of polymers is a relatively simple and more cost-effective method in comparison with other methods.

Until now, not only block copolymers of PLA and poly(ethylene glycol) (PEG) have been developed[8,9] but also miscibility and crystallization behaviors of PEG/PLLA blends have been paid much attention in many papers. It was reported that crystallization behavior of PLLA could be influenced in the existence of PEG[10-13]. In summary, PEG/PLA blends range from miscible to partially miscible, depending on composition[14-17].

In this study, PEG was used to make the blend with non-crystalline PLA (DLPLA) for promising improvement of the mechanical properties without lowering biodegradability, because PEG is also biodegradable. We examine effects of blending DLPLA with PEG on the higer-order crystalline structures of PEG by applying small-angle X-ray scattering (SAXS) measurements using high brilliant synchrotron radiation to detect nanostructures in the blends.

2. Experimental

The DLPLA sample was purchased from Sigma-Aldrich Co. and its weight-average molecular weight is in the range of 75,000 - 120,000. PEG with average molecular weight 20,000 was purchased from Wako Pure Chemical Industries, Ltd. (T_m = 65^{0}C measured by DSC). The use of the non-crystalline DLPLA sample enables us to examine the effect of blending on the higher-order crystalline structures of PEG, easier than for the case of the crystalline/crystalline blends.

All the blend samples were prepared by solution-casting method. Given amounts of DLPLA and PEG were dissolved in dichloromethane yielding ca. 5 wt% solution. The solution was poured into a Petri dish. A film was obtained after complete evaporation of dichloromethane under ambient condition at room temperature.

The SAXS and WAXS (Wide-angle X-ray scattering) measurements using the high brilliant synchrotron X-rays were carried out at BL-9C with the wavelengh of 1.5Å and BL-10C beamlines with wavelength of 1.488Å of the Photon Factory in the High Energy Accelerator Research Organization, Tsukuba, Japan. The imaging plate (250 x 250 mm^2), of which actual pixel size is 100 x 100 μm^2, was used as a two-dimensional (2d) detector. The typical exposure time was in the range 30-100s for SAXS and 10-30s for WAXS, respectively. BAS2000 (Fuji Photo Film Co., Ltd.) was used for development of exposed 2d images. The 2d-SAXS and 2d-WAXS patterns were further converted to one-dimensional profiles by conducting circular average and sector average.

3. Results and Discussion

Figure 1 shows the change in the one-dimensional SAXS (1d-SAXS) profiles measured at room temperature (around 25°C) for the PEG homopolymer (DL0) and various blend samples with different compositions. These profiles were obtained from two-dimensional SAXS (2d-SAXS) patterns by conducting the circular average, where q denotes the magnitude of the scattering vector as defined by $q = (4\pi/\lambda) \sin(\theta/2)$ with λ and θ being the wavelength of X-ray and the scattering angle, respectively.

For the PEG homopolymer (DL0) sample, multiple scattering peaks up to the fourth-order with the relative angular position of 1:2:3:4 are discernible. This suggests crystalline lamellae of PEG are regularly[18]. From the position q^* of the first-order peak (q^* = 0.264 nm^{-1}), the long period of the PEG crystalline lamellae was evaluated as 23.8 nm through the Bragg's equation $d = 2\pi/q^*$.

As for the blends, it is surprisingly found that the regularity became much better by blending 20 wt% of DLPLA (DL20), which can be concluded from the 1d-SAXS profile for the DL20 sample, exhibiting much sharper peaks than for the PEG homopolymer. Note here

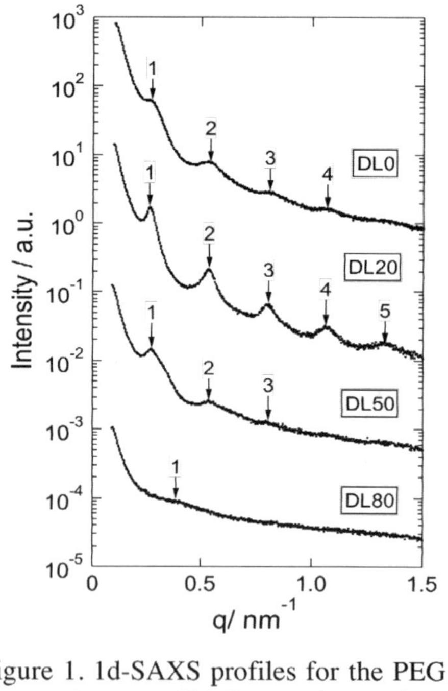

Figure 1. 1d-SAXS profiles for the PEG homopolymer (DL0) and various DLPLA/PEG blend samples (DL20, DL50, DL80) with different composition.

 IOP Publishing
doi:10.1088/1742-6596/272/1/012007

that the peak positions do not change, indicating no effect of blending on the lamellar repeating period. For the blending 50wt% of DLPLA (DL50), the peaks became broader as compared to the case of the homopolymer without changing the peak positions, as well. This means that the regularity of the lamellar stacking is spoiled by blending 50wt% of DLPLA without changing the repeating period. By further blending (80wt% of DLPLA; DL80), the 1d-SAXS profile is completely changed. One can find a broad peak at $q^* = 0.376\ nm^{-1}$, which gives the lamellar repeating period of 16.7 nm, 29.8% reduction of the period as compared to the homopolymer. The regularity of the lamellar repeating becomes very worse because the crystallinity of PEG is lowered by the large amount of amorphous DLPLA. The change in the repeating period with the blend composition is shown in Figure 2.

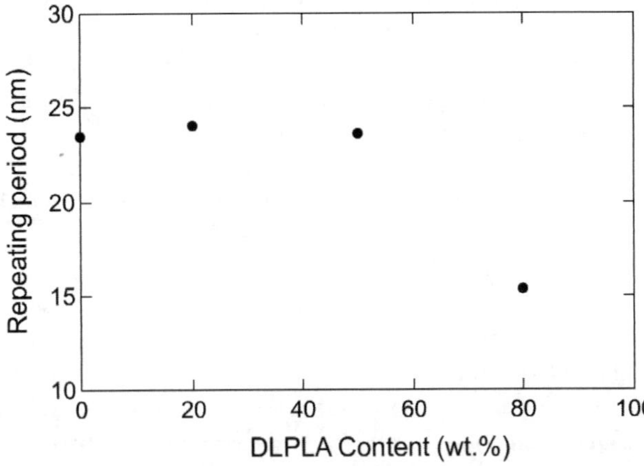

Figure 2. Change in lamellar repeating period with blended DLPLA content.

We are not sure why the 20wt% blending of DLPLA makes the PEG lamellar stacking much regular. One possible explanation may be due to the smooth interphase between crystalline and amorphous region for the blend than for the homopolymer because of more the regular chain folding. Also the change in the lamellar repeating period with the blend composition should be explained in the light of the miscibility between DLPLA and PEG in the amorphous phase. For this purpose, we further conducted the temperature dependence of the SAXS and WAXS measurements for the DL20 sample. The result is shown in Figures 3 and 4.

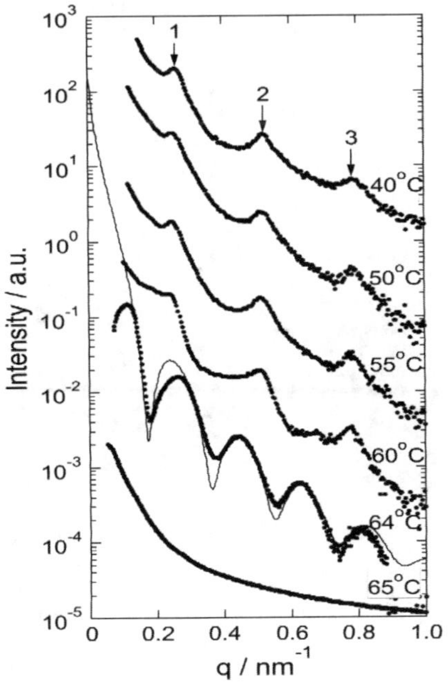

Figure 3. Change in the 1d-SAXS profile with temperature (heating process) for the DL20 sample.

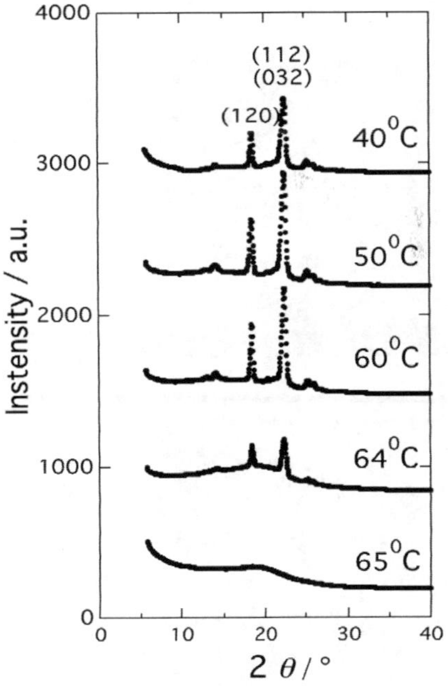

Figure 4. Change in 1d-WAXS profile with temperature (heating process) for the DL20 sample.

Though the SAXS and WAXS measurements were not conducted simultaneously, they can give supplementary information. Many higher-order SAXS peaks were observed for 40 ~ 55^{0}C due to the regular stacking of the crystalline. One can see the critical changes in the SAXS profiles at 64^{0}C and 65^{0}C. Since 65^{0}C is above the melting temperature of PEG, crystalline lamellae are subjected to melting. Therefore, no peak was observed both in the SAXS and the WAXS profiles (except for the amorphous peak in the WAXS profile). It is noticeable that the higher-order structure at only 1^{0}C below (at 64^{0}C) in the blend gives such a peculiar 1d-SAXS profile, which is completely different from the one measured at room temperature up to 60^{0}C in the heating process. The scattering comprises contributions from inter- and intra-particle interferences and those are referred to as lattice and particle scattering, respectively. This shape of the profile reminds us a scattering function for an isolated particle, we calculated model scattering function of an isolated lamellar particle[19].

From the best fitting between model calculation and the experimental curve, the thickness of the lamellar particle (L) was determined to be 36.5 nm. However, there is crucial deviation around q = 0.1 nm^{-1}. Especially, the model calculation cannot explain the peak at q = 0.13 nm^{-1} in the experimental profile. If it would be the first-order lattice peak, which gives the repeating period of 48.7 nm, the crystallinity estimated by L/D (L: lamellar thickness (36.5 nm) and D: lamellar repeating distance (48.7 nm)) would be 75% . This is rediculously large at temperature just 1^{0}C below T_m. As a matter of fact, by conducting peak decomposition of the WAXS profile measured at 64^{0}C, we evaluated the crystallinity 12%. Although it seems that the model with L = 36.5 nm and D = 48.7 nm may be non-realistic, there would be still a possible explanation if we consider that small grains, which contain a stack of the small number of the crystalline lamellae with thickness of 36.5 nm and the spacing of 48.7 nm, are sparsely dispersed in the matrix of completely amorphous phase.

In case that the peak around q = 0.13 nm^{-1} may not be the first-order lattice peak, we should improve modeling of the form factor. It should be noted that the model calculation we employed assumes the infinitesimal extent (width) of the plate-like particle. We should do calculation again with a finite size of the plate.

Figure 5 shows 2d-SAXS patterns for the through and edge view geometries of the as-cast sample (DL20). The edge view shows diffraction arcs (first- and second- order lattice peaks) with intensity

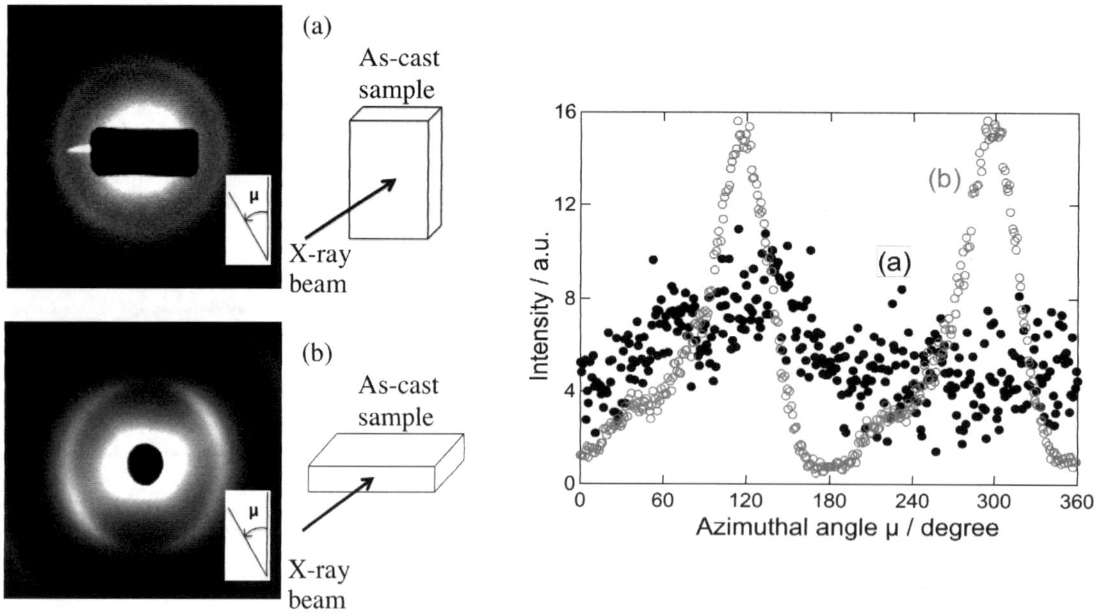

Figure 5. 2d-SAXS patterns of through view (a) and edge view (b) for the as-cast sample (DL20) at room temperature.

Journal of Physics: Conference Series **272** (2011) 012007 doi:10.1088/1742-6596/272/1/012007

accumulation in the equatorial direction, which is perpendicular to the normal of the as-cast film. This fact suggests that the crystalline lamellae are oriented more or less perpendicular to the surface of the as-cast film.

Why was such preferential perpendicular orientation spontaneously obtained? To explain this interesting phenomenon, temperature gradient due to the endothermic effect by the solvent evaporation during the solution cast may be a clue. To examine whether the PEG crystalline lamellae can be oriented by the temperature gradient, we have conducted separate experiments of thermal annealing of the PEG homopolymer under a horizontal temperature gradient and found that the crystalline lamellae were oriented parallel to the temperature gradient direction[20]. Therefore, this result can explain the spontaneous perpendicular orientation of the PEG lamellae in the as-cast film of the DLPLA/PEG blend sample due to the vertical temperature gradient induced by the solvent evaporation. Thus, further controls of the perpendicular orientation in the as-cast film promise to provide a material with special anisotropic properties by the simplest technique just like a solution casting[21].

To reveal higher-order structure formation from the melt upon cooling, we conducted SAXS measurements upon cooling from 65^{0}C. As shown in Fig. 6(a) neither particle nor lattice scattering appeared upon cooling from 65^{0}C. This result indicates that once the sample was melted away, the regular higher-order structure was not reformed. Here, it is noted that we confirmed crystallization by WAXS as shown in Fig. 6(b). Therefore, it is suggested that the regular crystalline lamellae (namely, the regular folding of the PEG chains) cannot be formed spontaneously although the crystallization of the PEG chains takes place. For further understanding of this strong hysteresis of the structural reformation, the SAXS measurement was conducted upon cooling from 64^{0}C, as shown in Fig. 6(c). Note that in this case we are intended to examine whether the regular lamellar stacking can take place or not upon cooling from 64^{0}C where we still observed the particle scattering as shown in Fig. 3. Since no lattice peak appeared, it is concluded that the lamellar stacking did not take place spontaneously, even in the case of cooling from the state where the regular isolated lamellae exist. Overall, we came conclusion that not only the chain folding but also the stacking of the isolated lamellae cannot take place spontaneously upon cooling. Taking into account of the fact that the lattice peaks were observed for the as-cast sample (Fig.1 & 3), the solution casting can be considered as a good technique for obtaining such regular higher-order crystalline structure. Finally, we should note that although the position of this peaks are slightly higher as compared to the case of the as-cast sample (Fig. 3), the lattice peaks were observed (Fig. 6a) after 8 days elapsed after the crystallization at room temperature from the melt (65^{0}C). Therefore, this result reminds that the formation of the regular higher-order crystalline structure takes comparatively long time.

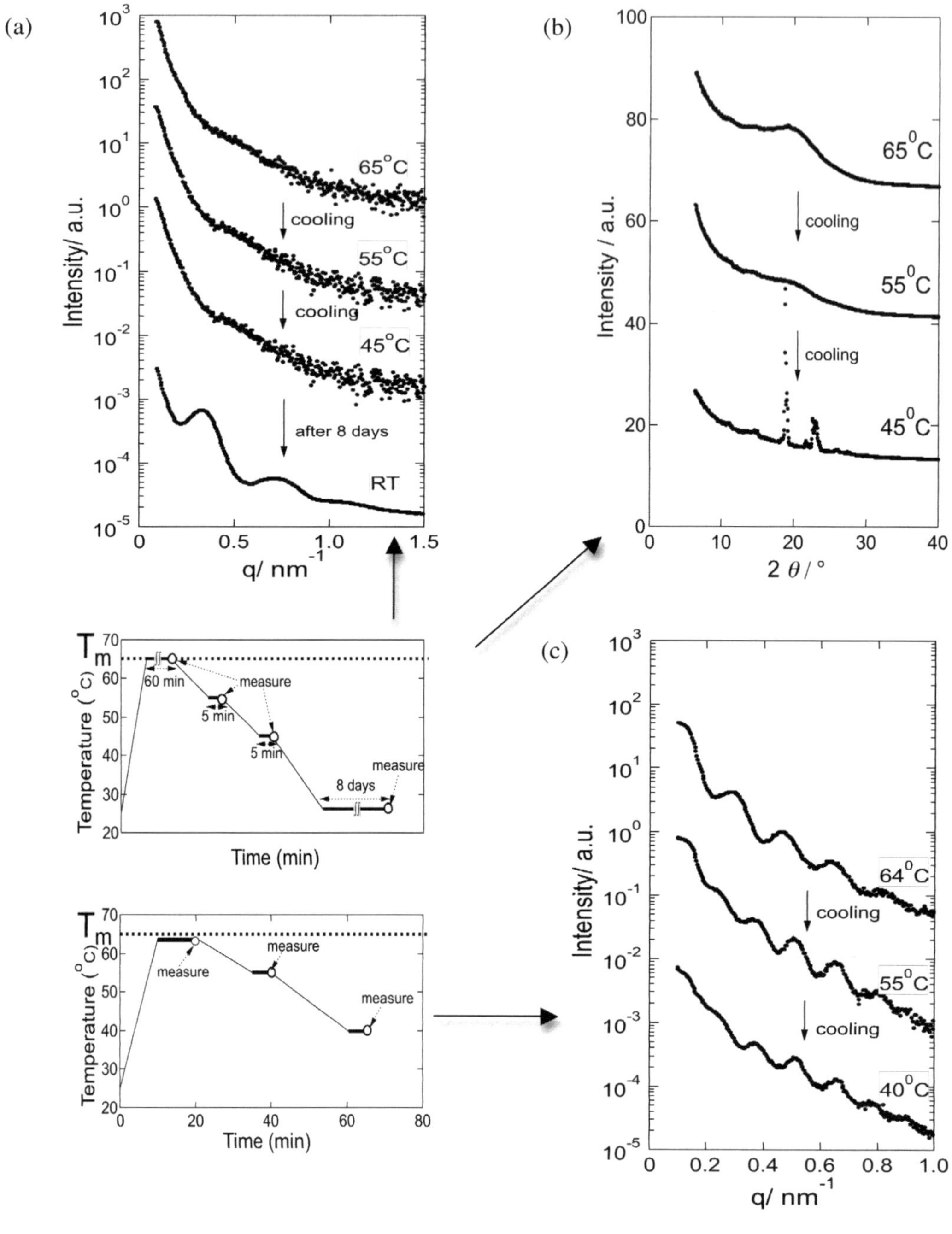

Figure 6. Change in the SAXS and WAXS profiles upon crystallization by cooling from 65°C (a & b) and from 64°C (c) for DL20 sample.

4. Conclusion

In this study, blends samples of non-crystalline DLPLA and PEG were subjected to structural analyses to examine effects of blending PLA (bio-based materials) with PEG on the higher-order crystalline structures of PEG. The structure is much more regular for blend (DLPLA/PEG = 20/80) than for the PEG homopolymer by observing many higher-order lattice peaks in the SAXS profile. Furthermore, for DLPLA/PEG = 20/80 blend sample, very perculiar SAXS profiles were observed just 1^0C below T_m of PEG. This is found to be a plate-like particle scattering, which has never been reported for crystalline polymers. In addition, it was found that there was strong hysteresis of the higher-order structure formation. Once the sample was melted away, the regular higher-order structure was not reformed. However, confirmed crystallization by WAXS suggests that the regular folding of the PEG chains cannot be formed spontaneously although the crystallization of the PEG chains takes place. In case of cooling from the state where the regular isolated lamellae exist, no lattice peak appeared. Therefore, not only the chain folding but also the stacking of the isolated lamellae cannot take place spontaneously upon cooling. Finally, the lattice (SAXS) peaks were observed after more than 1 day elapsed after the crystallization at room temperature from the melt (65^0C) reminds that the formation of the regular higher-order crystalline structure takes comparatively long time.

References

[1] Siparsky G L, Voorhees K J, Dorgan J R and Schilling K 1997 *J. Environ. Polym. Degrad.* **5** 125-36.
[2] Fukushima K, Hirata M and Kimura Y 2007 *Macromolecules* **40** 3049-55.
[3] Zu H, Yi C, Wang L and Xu W 2009 *Polym. Bull.* **64** 471-81.
[4] Jamshidi K, Hyona S H and Ikadaa Y 1988 *Polymer* **29** 2229-34.
[5] Meaurio E, Zuza E and Sarasua J R 2005 *Macromolecules* **38** 1207-15.
[6] Wang H, Zhang Y, Tian M, Zhai L, Wei Z and Shi T 2008 *J. Appl. Polym. Sci.* **110** 3985-89.
[7] Ray S S, Yamada K, Okamoto M, Fujimoto Y, Ogami A and Ueda K 2003 *Polymer* **44** 6633-46.
[8] Fujiwara T, Miyamoto M and Kimura Y 2001 *Macromolecules* **34** 4043-50.
[9] Fujiwara T, Miyamoto M, Kimura Y and Sakurai S 2001 *Polymer* **42** 1515-23.
[10] Younes H and Cohn D 1988 *Eur. Polym. J.* **24** 765-73.
[11] Nakafuku C and Sakoda M 1993 *Polym. J.* **25** 909-17.
[12] Nakafuku C 1994 *Polym. J.* **26** 680-87.
[13] Nakafuku C 1996 *Polym. J.* **28** 568-75.
[14] Lai W C, Liau W B and Lin T T 2004 *Polymer* **45** 3073-80.
[15] Sheth M, Kumar R A, Dave V, Gross R A and McCarthy S P 1997 *J. Appl. Polym. Sci.* **66** 1495-1505.
[16] Nijenhuis A J, Colstee E, Grijpma D W and Pennings A J 1996 *Polymer* **37** 5849-57.
[17] Jang J M, Chen H L, You J W and Hwang J C 1997 *Polym. J.* **29** 657-62.
[18] Shiomi T, Takeshita H, Kawaguchi H, Nagai M, Takenaka K and Miya M 2002 *Macromolecules* **35** 8056-65.
[19] Sakurai S, Okamoto S, Kawamura T and Hashimoto T 1991 *J. Appl. Cryst.* **24** 679-84.
[20] Kimura G, Yasuda, Sasaki S, Sakurai S to be submitted.
[21] Sakurai S, Bando H, Yoshida H, Fukuoka R, Mouri M, Yamamoto K and Okamoto S 2009 *Macromolecules* **42** 2115-21.

Journal of Physics: Conference Series **272** (2011) 012008 doi:10.1088/1742-6596/272/1/012008

Lamellar-lamellar phase separation of phospholipid bilayers induced by salting-in/-out effects

Mafumi Hishida[1], **Hideki Seto**[2*]
[1] Institute for Integrated Cell-Material Sciences, Kyoto University, Kyoto 606-8501, Japan
[2] KENS & CMRC, Institute of Materials Structure Science, High Energy Accelerator Research Organization, Tsukuba 305-0801, Japan

[*]E-mail: hideki.seto@kek.jp

Abstract. The multilamellar structure of phospholipid bilayers is stabilized by the interactions between bilayers. Although the lamellar repeat distance is uniquely determined at the balance point of interactions between bilayers, a lamellar-lamellar phase separation, where the two phases with different lamellar repeat distance coexist, has been reported in a case of adding a salt to the aqueous solution of lipids. In order to understand the physical mechanism of the lamellar-lamellar phase separation, the effects of adding monovalent salt on the lamellar structure are studied by visual observation and by small-angle X-ray scattering. Further, a theoretical model based on the mean field theory is introduced and it is concluded that the salting-in and -out effects of lipid bilayers trigger the lamellar-lamellar phase separation.

1. Introduction

The phospholipid bilayer is the basic structure of biomembranes. Through the simple mixing of synthesized phospholipids and water, the lipids spontaneously form bilayers and are organized as multilamellar vesicles, where a number of bilayers stack regularly. The characteristic repeat distance between bilayers is uniquely determined by the balance of attractive and repulsive interactions. Four main interactions have been considered; van der Waals attraction, hydration repulsion, steric repulsion caused by the thermal fluctuation of bilayers and electrostatic interaction [1]. In cases of neutral lipids, the electrostatic interaction can be ignored. These interactions equilibrate the multilayer repeat distance at 5 to 10 nm. On the other hand, Rappolt et al. observed that two or three lamellar structures with different repeat distances coexist when monovalent salt is added to the aqueous solution of neutral phospholipid in the liquid-crystalline L_α phase [2,3]. This evidence contradicts the model to explain the lamellar repeat distance explained above, and the mechanism of the lamellar-lamellar phase separation has not yet been clarified, even though the authors suggested that osmotic stress is strongly related to the phase separation.

With respect to the effects of monovalent salt on the phospholipid lamellar phase, several intriguing results have been reported in relation to the 'salting-out effect', i.e., solutes aggregate in high salinity by a kind of repulsive force between salt and solute, and the 'Hofmeister series', i.e., the ion specificity of the salting-out effect [4-9]. Petrache et al. observed neutral phospholipid suspension,

Journal of Physics: Conference Series **272** (2011) 012008 doi:10.1088/1742-6596/272/1/012008

including KCl or KBr, by visual inspection, and it was revealed that multilamellar vesicles float in the salt solution [6]. This high buoyancy of vesicles indicates that a kind of repulsive force exists between phospholipid bilayers and salt ions, and it changes the relation between specific gravities of multilamellar vesicles and bulk brine. This force could be originated by the salting-out effect, and the strength differences of the repulsive force between KCl and KBr solutions follow the 'Hofmeister series' [4,5]. The ion specificity is additionally reported in terms of the Debye screening of the van der Waals attraction [7,8]. Furthermore, Korreman et al. investigated the effect of monovalent salt on the anomalous swelling of multilamellar vesicles [9]; the increase of the lamellar repeat distance when approaching the main transition temperature from the L_α to L_β phase [10-12]. They demonstrated that the anomalous swelling is amplified by increasing the concentration of monovalent salt, and the anomalous swelling is explained as a critical unbinding with the critical exponent 1. The behaviour of anomalous swelling also followed the Hofmeister series, and the order of the effect of adding monovalent salts is as follows: NaCl < KF < KCl < NaF < NaBr < KBr.

Theoretically, the lamellar-lamellar phase separation of neutral lipid membrane has been discussed in relation to the pre-unbinding of stacked membranes [13,14]. The mean field theory of pre-unbinding has been constructed with the Flory theory of multilayers [13]. In the theory, the free energy is described as

$$g(\phi) = -\chi\phi^2 + \frac{\phi^3}{2(1-\phi)^2} - \mu\phi,$$

where the order parameter ϕ is the volume fraction of lipid [15]. The first term indicates the direct interaction between membranes, the second one is the entropic term of stacked lipid membranes, and the third one is for conservation of the total volume fraction. In the case of $\chi > 0$, the lamellar phase coexists with bulk water. In the theory of pre-unbinding, a secondary order parameter other than the lipid concentration is applied, and the lamellar phase separation to two or three phases is induced in the vicinity of the critical unbinding point [13]. In the theory, salt concentration possibly become the second-order parameter and could be the origin of the lamellar phase separation.

In the present study, we discuss the salting-out effect on the lamellar-lamellar phase separation. As a result, a different salt concentration between the two separated lamellar phases is observed, and the salting-in and -out effects are considered to be a main cause of the lamellar-lamellar phase separation. Furthermore, we suggest a theoretical model of the lamellar-lamellar phase separation with considering salting-in/-out effects, based on the pre-unbinding theory. These results may assist in understanding the salting-out effect and the Hofmeister series, which remains a matter of dispute.

2. Materials and methods

As a neutral phospholipid, 1,2-dipalmitoyl-sn-glycero-3-phosphocholine (DPPC, Wako Pure Chemical Industries) was obtained in powder form and used without further purification. Sodium chloride (NaCl, Wako Pure Chemical Industries) was first dissolved in pure water at several concentrations. Next, the powder DPPC was mixed in the solutions to be 8 wt% or 20 wt% at above the main transition temperature. The salt concentrations were [DPPC]:[NaCl](mol/mol) = 1:0 (Pure DPPC), 1000:1, 100:1, 10:1, 1:1, and 1:10. Because NaCl is categorized as "water-structure-making" salt, it is expected to exhibit a high salting-out effect. We also used NaBr and KBr as a salt that exhibit less salting-out effect.

For measuring the repeat distances of the multilamellar phase of the samples, small angle X-ray scattering (SAXS) was performed at BL-15A, Photon Factory, KEK, Japan. Supplemental experiments were performed with SAXS in the laboratory (RA-Micro7HF, Rigaku). A sharp Bragg peak at q observed by SAXS corresponds to the characteristic repeat distance of lamellar phase $d = 2\pi/q$. Because the main transition temperature of DPPC is approximately 42°C, the anomalous swelling was exhibited 42°C < T < 50°C [11]. The buoyancies of multilamellar vesicles were observed by visual observation to evaluate the salting-out effect.

3. Results

Figure 1 (a) presents a typical image of the macroscopic phase separation of multilamellar vesicles and bulk water in NaCl solution at 50°C. In pure water, the L_α lamellar phase of DPPC sinks slightly (the specific volumes are 1.009 mL/g for DPPC and 1.012 mL/g for H_2O at 50°C [16]). As we add a little amount of salt ([DPPC]:[NaCl] = 1000:1, 100:1, 10:1), the multilamellar vesicles become heavier than the bulk water (see from 1:0 to 100:1). However, when the ratio of NaCl is more than 1:1, the balance of specific gravities between the multilamellar vesicles and bulk water is reversed, and the lamellar phase floats on the upper part of the test tube at high salt concentration.

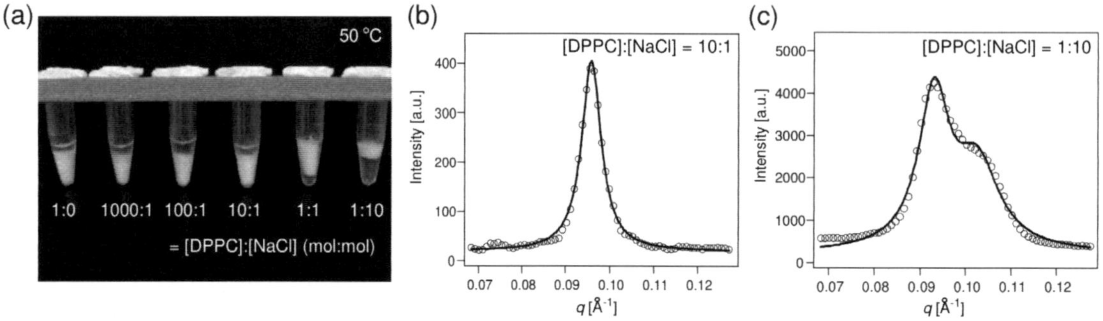

Figure 1. (a) Picture of 8 wt% DPPC solutions with different NaCl concentrations at 50°C. The white suspension is phospholipid multilamellar vesicles. (b)(c) SAXS profiles of [DPPC]:[NaCl] = 10:1 and [DPPC]:[NaCl] = 1:10 solutions at 50°C, respectively. The solid lines indicate fitting results by Lorentzian and double-Lorentzian functions, respectively.

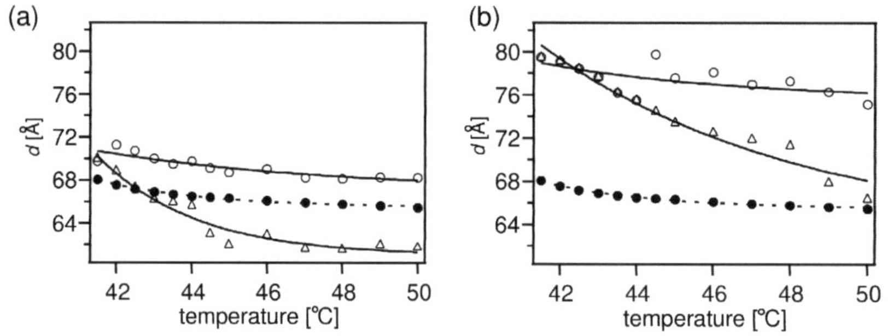

Figure 2. Anomalous swelling behaviour of pure DPPC and DPPC/NaCl solutions ([DPPC]:[NaCl] = 1:10, (a)[DPPC] = 8 wt% and (b) [DPPC] = 20 wt%). Closed circles denote the lamellar repeat distance of pure DPPC, whereas open circles and open triangles are the phase-separated lamellar repeat distances of DPPC/NaCl solutions. Solid and dashed lines are the fitting results by an exponential function.

The specific gravity of the phospholipid bilayer should be identical even with salt because the bilayer structure is the same [7]. Thus, the observed behaviour may be caused by the asymmetric distribution of the salt between the interlamellar water and the bulk water. When a small amount of salt is dissolved in the solution, the salt is concentrated in the interlamellar water and the multilamellar vesicles sink: the "salting-in effect." However, as the amount of salt is increased, the local concentration of salt in bulk water become higher than that in the interlamellar water, and the vesicles

Journal of Physics: Conference Series **272** (2011) 012008 doi:10.1088/1742-6596/272/1/012008

float as a result of buoyancy: "salting-out effect." In the system of DPPC and NaCl at 50°C, the criteria from salting-in to salting-out is determined between [DPPC]/[NaCl] = 10 and 1.

By performing SAXS experiments for these samples at 50°C, we observe the salt concentration dependence of the lamellar-lamellar phase separation. Whereas a single Bragg peak is exhibited under the salt concentration below [DPPC]/[NaCl] = 1 (figure 1 (b)), evident phase separation is observed as salt is concentrated (figure 1 (c)). The temperature dependence of the repeat distances of the coexisting phases at [DPPC]/[NaCl] = 1/10 [mol/mol] just above the main transition temperature (the anomalous swelling region) are depicted in figure 2. The two separated phases exhibit different anomalous swelling behaviours for both samples of [DPPC] = 8 wt%, 20 wt%: the lamellar phase with a larger repeat distance swells slightly as it approaches the transition temperature (42°C), whereas that with a smaller repeat distance swells significantly on cooling. This asymmetric behavior of anomalous swelling could be caused by the heterogeneous distribution of salt ions between the two separated lamellar phases, because the degree of the anomalous swelling depends on the salt concentration between layers [9]. However, we were unable to determine distribution of salt, because our results exhibit an opposite tendency to those of Korreman [9].

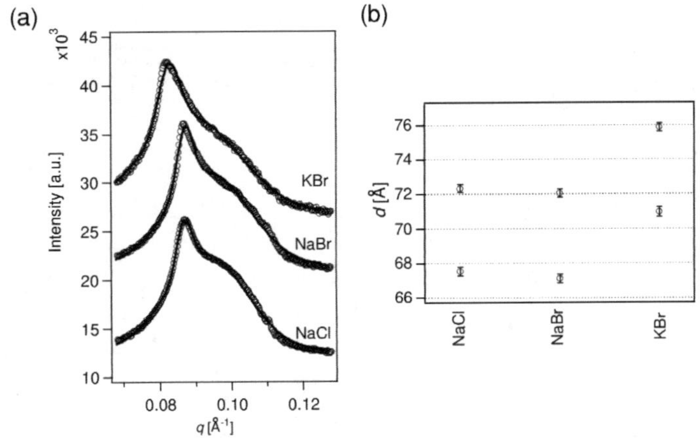

Figure 3. (a) SAXS profiles of DPPC with salts; NaCl, NaBr, KBr. Salt concentrations were [DPPC]:[Salt]= 1:10 (mol:mol), and DPPC concentrations were 8 wt%. The solid lines are the fitting results by double-Lorentzian functions. (b) Each repeat distance of phase-separated lamellae.

From these results, the origins of the lamellar-lamellar phase separation can be considered as a result of the reduction of the van der Waals attraction due to the Debye screening and salting-in and -out effects. The dependence of ion species on the lamellar-lamellar phase separation confirms the salting-in and -out effects as an essential origin of the lamellar phase separation (figure 3). According to the Hofmeister series, the ordering of salt for salting-in and -out effects is NaCl < NaBr < KBr. SAXS results indicate that the solutions with NaCl, NaBr, and KBr exhibit different lamellar phase separation behaviour, even when the salt concentrations are exactly the same. The exchange of a potential of hydration repulsion by monovalent salt ions is also considered, because the "water-structure-making" ions strongly bind water molecules [4] and the water layer on the head group of phospholipids would be broken up. However, the hydration repulsion acts as a very short-range interaction, ranging approximately 2 Å [1,17], whereas our results regarding different phase separations between NaCl and KBr solutions range up to 4 Å. This result is inconsistent with the possibility that the exchange of hydration repulsion is an essential origin of the lamellar-lamellar phase separation.

4. Discussion

Journal of Physics: Conference Series **272** (2011) 012008 doi:10.1088/1742-6596/272/1/012008

Based on the Flory theory of pre-unbinding [13], we discuss the lamellar-lamellar phase separation, considering the salting-in/-out effects and the Debye screening of the van der Waals interaction. In the case of the lamellar phase with salt ions, the concentration of salt is defined as the second-order parameter ψ in the mean field theory. Four contributions of salt ions to the free energy are considered: Debye screening of the van der Waals interaction, salting-in/-out effects, translational entropy of salt, and salt concentration. Thus, the free energy is described as

$$g(\phi,\psi) = -(\chi_0 + B\psi)\phi^2 + \frac{\phi^3}{2(1-\phi)^2} - \mu\phi + \psi\log\psi + \alpha(\psi - \frac{\beta}{\alpha})\phi - \tilde{\mu}\psi \cdot$$

The first term indicates the direct interaction between lipid bilayers with the Debye screening of the van der Waals interaction caused by the effect of salt ($B \le 0$), the fourth term is the translational entropy of salt ions, the fifth term reflects the salting-in ($\psi < \beta/\alpha$) and salting-out ($\psi > \beta/\alpha$) effects, and the third and sixth terms denote the conservation of lipid and salt. By minimizing the free energy with respect to ψ

$$\frac{\partial g}{\partial \psi} = 0,$$

the concentration of salt is calculated as

$$\psi_m = \exp(B\phi^2 - \alpha\phi + \tilde{\mu} - 1).$$

The substitution of ψ_m to the free energy gives the free energy with respect to the concentration of lipid membrane, $g(\phi)$.

In figure 4 (a), the typical free energy profile with respect to the concentration of lipid membrane is shown. $\phi = 0$ denotes the bulk water phase. In a certain parameter range, three local munima at $\phi = \phi_1, \phi_2, 0$ are observed, which indicates that two separated lamellar phase coexist with bulk water. The salt concentration is found to distribute heterogeneously (figure 4 (b)), as observed in our experiments. This theory suggests that the key factor for the lamellar-lamellar phase separation is not the Debye screening of the van der Waals interaction but the salting-in/-out effect, because the free energy exhibits a double minimum profile even if $B = 0$.

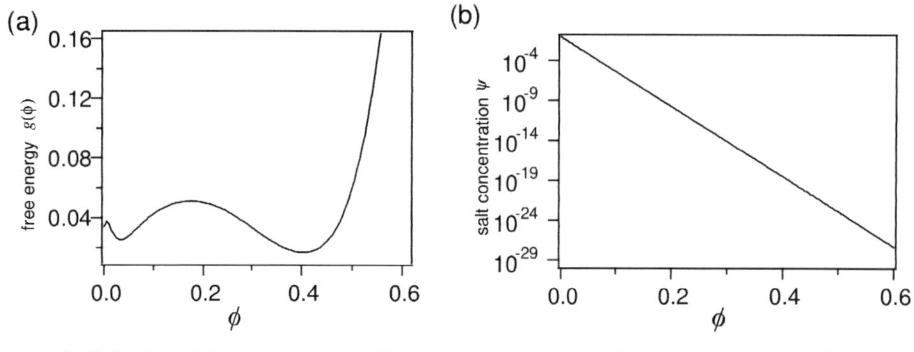

Figure 4. Calculated free energy profile (a) and corresponding salt concentration (b) with respect to the lipid concentration ϕ at $\chi = 2$, $\mu = -2$, $B = -1$, $\tilde{\mu} = -1.5$, $\alpha = 100$, $\beta = 1.38$.

5. Conclusion

In this study, we investigated the physical mechanism of the lamellar-lamellar phase separation by adding monovalent salt. The visual observation of the buoyancy of lipid vesicles and SAXS measurements of the anomalous swelling behaviour indicate that salt concentrations are different among each of the separated lamellar phases and the bulk water phase. As a result, it is clarified that a certain interaction exists between lipid bilayers and salt ions, which becomes both attractive and repulsive depending on the salt concentration, i.e., salting-in/-out effects. The SAXS measurements also indicate that the lamellar phase separation differs by ion specificity. These experimental results suggest that salting-in/-out effects dominate the lamellar phase separation accompanied with the Debye screening of the van der Waals interaction by salt. From the theoretical model based on the mean field Flory theory of the pre-unbinding with consideration of the salting-in/-out effects and the Debye screening of the van der Waals interaction, it is concluded that the salting-in/-out effects are the essential origin of the ion-induced lamellar-lamellar phase separation.

Acknowledgements
The authors thank Professor K. Yoshikawa (Kyoto University), N. L. Yamada (KEK), and Dr. H. Mayama (Hokkaido University) for the helpful discussions and experimental support. The SAXS experiments were performed under the approval of the Photon Factory, KEK (No. 2006G064). M.H. was supported by a Grant-in-Aid for JSPS Fellows (No. 19-2774) and a start-up fund provided by Kyoto University.

[1] Petrache H I, Gouliaev N, Tristram-Nagle S, Zhang R, Suter R M, Nagle J F 1998 *Phys. Rev. E* **57** 7014
[2] Rappolt M, Pressl K, Pabst G, Laggner P 1998 *Biochim. Biophys. Acta* **1372** 389
[3] Rappolt M, Pabst G, Amenitsch H, Laggner P 2001 *Colloid. Surf. A* **183-185** 171
[4] Zhang Y, Cremer P S 2006 *Curr. Opin. Chem. Biol.* **10** 658
[5] Boström M, Wiliams D R M, Stewart P R, Ninham B W 2003 *Phys. Rev. E* **68** 041902
[6] Petrache H I, Kimchi I, Harries D, Parsegian V A 2005 *J. Am. Chem. Soc.* **127** 11546
[7] Petrache H I, Tristram-Nagle S, Harries D, Kucerka N, Nagle J F, Parsegian V A 2006 *J. Lipid Res.* **47** 302
[8] Petrache H I, Zemb T, Belloni L, Parsegian V A 2006 *Proc. Natl. Acad. Sci. USA* **103** 7982
[9] Korreman S S, Posselt D 2000 *Eur. Biophys. J.* **30** 121
[10] Chu N, Kucerka N, Liu Y, Tristram-Nagle S, Nagle J F 2005 *Phys. Rev. E* **71** 041904
[11] Pabst G, Amenitsch H, Kharakoz D P, Laggner P, Rappolt M 2004 *Phys. Rev. E* **70** 021908
[12] Seto H, Yamada N L, Nagao M, Hishida M, Takeda T 2008 *Eur. Phys. J. E* **26** 217
[13] Komura S, Shimokawa N, Kato T 2006 *J. Chem. Phys.* **124** 034906
[14] Philipp K, Schick M 1996 *Phys. Rev. E* **54** R33
[15] Milner S T, Roux D 1992 *J. Phys. I France* **2** 1741
[16] Wiener M C, Tristram-Nagle T, Wilkinson D A, Campbell L E, Nagle J F 1988 *Biochim. Biophys. Acta* **938** 135
[17] Milhaud J 2004 *Biochim. Biophys. Acta* **1663** 19

Journal of Physics: Conference Series **272** (2011) 012009 doi:10.1088/1742-6596/272/1/012009

A scanning SAXS/WAXS study of rat brain

Naoto Yagi[1,2,3]

[1]Japan Synchrotron Radiation Research Institute, 1-1-1 Kouto, Sayo, Hyogo 679-5198, Japan.
[2]RIKEN SPring-8 Center, 1-1-1 Kouto, Sayo, Hyogo 679-5148, Japan.
[3]CREST, Japan Science and Technology Agency, Kawaguchi, Saitama 332-0012, Japan.

E-mail: yagi@spring8.or.jp

Abstract. A simultaneous SAXS (small-angle X-ray scattering) and WAXS (wide-angle X-ray scattering) measurement setup was installed at BL45XU in SPring-8. The system comprises of a short (specimen-to-sample distance about 50cm) vacuum path and a mosaic CCD detector. It covers a q-range of 0.02-2.5 nm^{-1}. Using this setup, lipids in formalin-fixed rat brain were analyzed. A brain slice was moved across the X-ray beam with a step size of 0.5 mm to map reflections from lipids in various areas of brain. White matter that contains myelin gave strong lamellar reflections in the small-angle region which are often unisotropic. Gray matter shows only a central scatter in the small-angle region. In the wide angle region, both white and gray matters gave rise to sharp rings that are due to lateral packing of hydrocarbon chains in the lipid membranes. The relative intensities of these rings were different in white and gray matters, showing that the lateral arrangements of the lipids in bilayers are different.

1. Introduction

Most of biological tissues have a hierarchical structure that comprises of distinct units at different levels of scale. Since the structures at different scales give X-ray diffraction or scattering in different angular ranges, it is often useful to record the small- and wide-angle diffraction from the same sample to understand its structure. At synchrotron radiation facilities, this has been realized by using separate detectors for the small- and wide-angle regions [1,2].

Brain is a typical example of a hierarchical structure. Brain is mostly made of neuronal cells that are rich in lipids. Different regions of brain are specified by their appearance (color) such as white, gray and dark matters. The major reason for the different appearances is a difference in the lipid structure and composition. Gray matter is mostly made of cell bodies and neurons while white matter contains axons that are shielded by myelin. Myelin is produced by Schwann cells and structurally a stack of double bilayers of lipids. Myelin appears white, providing the origin of the color of white matter. The chemical compositions and the structure of lipids in brain are strongly related to its medical conditions. It has been reported that lipid modifications in gliomas, the most frequent primary brain tumors, correlate with their malignancy [3]. It has been also shown that, unlike normal myelin, multiple sclerosis myelin does not show a 4.15nm peak that represents gel phase, possibly as a result of different protein-lipid interaction [4]. Myelin in brain tumors has been studied by X-ray diffraction recently [5-7], but the functional relation between its lipid constituents and brain activity is not well understood.

In the present study, slices of rat cerebrum were subjected to a scanning SAXS/WAXS measurement to study the lipid structure. The experimental setup is based on a use of a large CCD detector for protein crystallography. This technique allows us to make a two-dimensional distribution map of materials in a sample. This is extremely useful for a specimen like brain in which different constituents are mixed but segregated in some regions.

2. Materials and Methods

2.1. Preparation of brain samples

Cerebrum was excised from male/female Wister or Sprague-Dawley rats that were used in experiments at SPring-8 (mostly for the studies on cardiac muscle). After the experiments, the rats were killed by an overdose of pentobarbital and the brains were dissected out. They were put into neutral-buffered 10% formalin and kept in a refrigerator for one week to three months. No difference was observed with the preservation period nor when the sample was preserved at room temperature
The formalin-fixed brains were sliced with a razor to a thickness of 1-2 mm for 2D scanning. The slices were kept in a phosphate-buffered saline (pH 7.0) and sealed between two sheets of Mylar (thickness 6 μm) with the saline. X-ray measurements were made at room temperature.

2.2. Simultaneous WAXRD and SAXS measurement at BL45XU

BL45XU has two tandem helical undulators as an X-ray source and uses a diamond crystal as a beam splitter [8,9]. In the present experiment, the protein crystallography hutch with three sets of diamond double-crystal monochromators was used. One set of the monochromators was tuned to obtain a 13.8 keV X-ray, while the others were moved out of the beam. A bent cylindrical mirror was used to focus the X-ray beam. The beam size was approximately 250 μm × 250μm at the sample. The sample-to-detector distance was 450 mm. At 15 mm upstream to the specimen was a guard pinhole with a diameter of 0.8 mm (Figure 1a). The flux was about 2×10^{11} cps. The X-ray detector was Jupiter210 (RIGAKU, Tokyo, Japan), which comprises of four CCDs with tapered fibers. The detection area was 205 mm × 205 mm. The pixel size was 100 μm × 100 μm with 2×2 binning. In order to cover a wide q-range, the detector was set at an offset position so that the beam stop was near the bottom of the detection area (Figure 1b). The SAXS and WAXS regions were measured simultaneously. A fiber diffraction pattern from dried chicken tendon collagen (Figure 1c) shows the 3rd-order meridional reflection at a Bragg spacing of about 1/21 nm^{-1} and the meridional reflection at 1/0.28 nm^{-1} (Figure 1b), demonstrating that this setup covers a q-range of about two orders of magnitude ($q=4\pi\sin(\theta)/\lambda$, the Bragg spacing $d=2\pi/q$, where 2θ is the scattering angle and λ is the wavelength). Since the SAXS intensity exceeds the WAXS intensity by nearly three orders of magnitude and the dynamic range of the detector was several thousand, the software for the measurement control was designed to allow multiple exposures with different exposure time. Thus, to measure the wide-angle region with good statistics and to avoid saturation in the small-angle region, images with three different exposure times (typically 2, 5 and 30 sec) were recorded from the same spot of the sample. The dark images (images taken with the same exposure time without X-rays) was subtracted from each image. Also, a small (< 5%) correction was made to cancel the differences in amplification of the four CCDs. For intensity measurements, a diffraction pattern was averaged circularly to obtain a one-dimensional intensity profile. Then, the integrated intensities of peaks were obtained by fitting a polynomial (usually a second order) background and Gaussian peaks whose center positions were used as the peak positions.

3. Results

3.1 Small-angle X-ray scattering (SAXS)

The SAXS patterns from rat brain differ markedly in regions. In gray matter of cerebrum that includes cerebral cortex and caudate nuclei, only diffuse scatter was observed in the small-angle region. On the other hand, in white matter including external capsule, strong rings or arcs were observed showing the abundance of myelin sheaths (Figure 2a). These can be indexed on a fundamental repeat of 14.4 nm.

This is shorter than the repeat distance in unfixed brain (15.9 nm) and other reports (18.0 nm in native rabbit sciatic nerve, 15.6 nm in rabbit optic nerve, 17.0 nm in frog sciatic nerve myelin [10]. Most probably, formalin reduced the distance between lipid bilayers by cross-linking the proteins that are located outside of the bilayer. From formalin-fixed human cerebral myelin, De Felici et al.[5] observed a 16.0-16.6nm periodicity, but the length of fixation was only a few days in their experiments while it was longer than one week in the present case. The myelin reflections were usually observed up to the sixth order and often showed orientation indicating the direction of the axons. There were also other weak rings of unknown origin such as broad maxima at d=1.1 and 0.72 nm. As for the former, Inouye and Kirschner [11] observed a weak diffuse ring at d=1.1 nm in central myelinated tissue which they attributed to myelin protein.

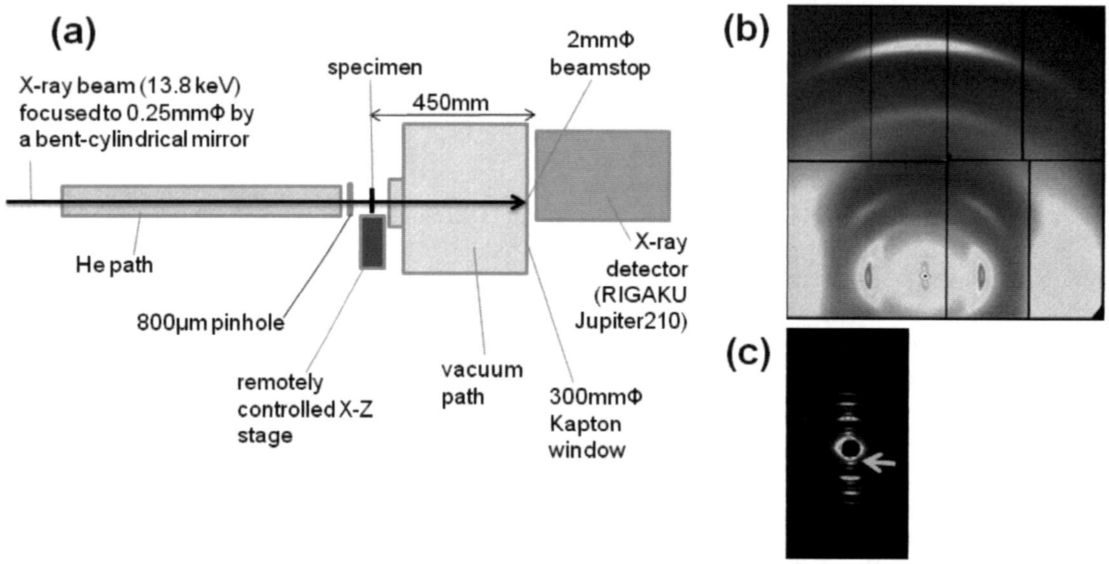

Figure 1. The SAXS/WAXD measurement setup at the BL45XU protein crystallography hutch in SPring-8. (a) Arrangement in the experimental hutch. The area detector (Jupiter210, RIGAKU, Tokyo, Japan) was placed off-center to cover a large q-range. (b) A wide-angle diffraction pattern from collagen (dried chicken tendon). The strong arc at the top is the 0.28nm meridional reflection. The black lines are unusable regions due to gaps between tapered fibers and to boundary between segments of the CCDs. (c) A small-angle diffraction pattern from collagen. The peak just outside of the beam stop (pointed by an arrow) is the third order meridional reflection from the 63nm periodicity.

3.2 Wide angle X-ray diffraction (WAXD)

The WAXD patterns from formalin-fixed brain samples showed well-defined peaks in both gray and white matters (Figure 2b), which can be attributed to lateral packing of lipid molecules in bilayers. There were several peaks (Figure 3): a weak, broad peak at q=14.26 nm^{-1}, a weak, sharp peak at 14.68 nm^{-1}, a small shoulder at 14.81 nm^{-1}, a well-defined strong peak at 14.93 nm^{-1}, another well-defined strong peak at 15.27 nm^{-1}, a broad, weak shoulder at 15.48 nm^{-1} and a broad peak at 16.57 nm^{-1}. There were other rings that were not observed constantly. Slices of brains not fixed with formalin did not show sharp peaks in white or gray matter. Occasionally, a broad peak was observed around q=15 nm^{-1}. The intensity of the peaks varied according to the areas of brain. Especially, the relative intensity of the two strong reflections at 14.93 nm^{-1} (d=0.421nm) and 15.27 nm^{-1} (d=0.411nm) reversed between

Journal of Physics: Conference Series **272** (2011) 012009 doi:10.1088/1742-6596/272/1/012009

white and gray matters (Figure 3). From formalin-fixed human cerebral myelin, De Felici et al. [5] observed a peak at q=14.4 nm^{-1} (d=0.436nm) that may correspond to the former peak but much broader than in the current experiment.

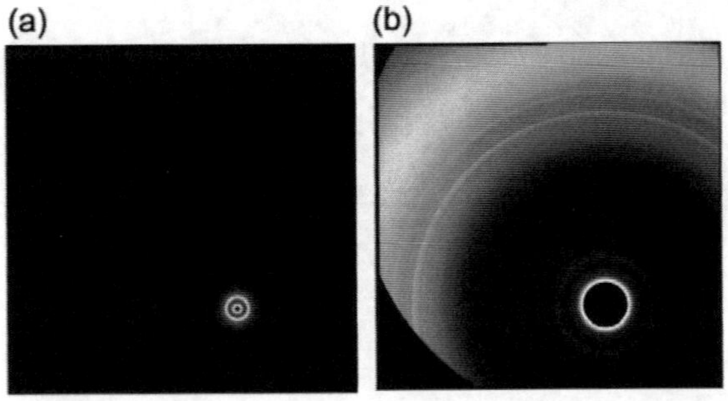

Figure 2. The SAXS/WAXD recording from formalin-fixed white matter (myelin) of rat cerebrum. (a) Small-angle region recorded with an exposure time of 2 sec, showing the strong second order peak from myelin. Although not apparent in this pattern, there were higher orders which were often oriented and appeared as arcs. (b) Wide-angle region recorded with a 30sec exposure showing sharp rings from lipids and a broad ring from water. The small-angle region is saturated. In these images, after dark subtraction, the gaps between the tapered fibers and other defects were filled by interpolation.

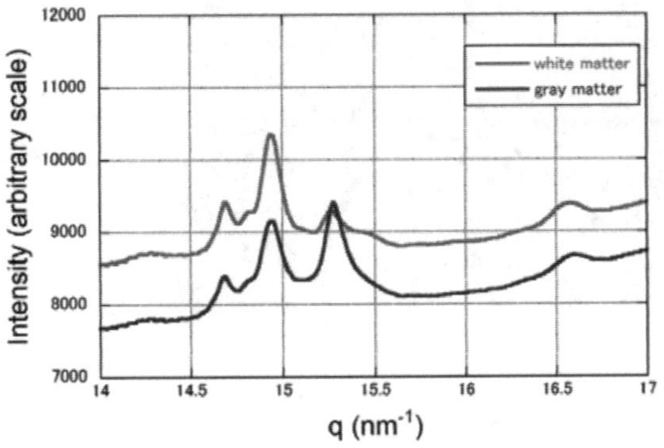

Figure 3. The wide-angle intensity profiles observed from rat cerebrum. Profiles obtained in two regions, white and gray matters, are shown. The high background is from other components in the specimen such as proteins and water.

3.3 2D mapping of the SAXS and WAXD

To investigate the relative intensity of the lipid-derived peaks, diffraction patterns were obtained from different regions of cerebrum by moving the sample with 0.5mm steps in two dimensions. Figure 4 shows maps of intensity of the myelin 4th-order reflection and peaks at 14.93 and 15.27 nm^{-1}. The myelin reflection was generally strongest in white matter. Since the slice is 1-2mm thick, there may be white matter that cannot be seen in the photograph shown in Figure 4a. This may have somewhat blurred the map. The distribution of intensity of the myelin reflection (Figure 4b) is similar to that of the wide-angle reflection at 14.93 nm^{-1} (d=0.421nm) (Figure 4c). On the other hand, the reflection at 15.27 nm^{-1} (d=0.411nm) is stronger in the cortex and the central part of cerebrum (Figure 4d). Although the two distributions are not mutually exclusive, it is clear that white and dark matters are characterized by these reflections.

The distribution of the intensity of the reflections at 14.68 nm^{-1} and 16.57 nm^{-1} also correlated weakly with that of the myelin reflection.

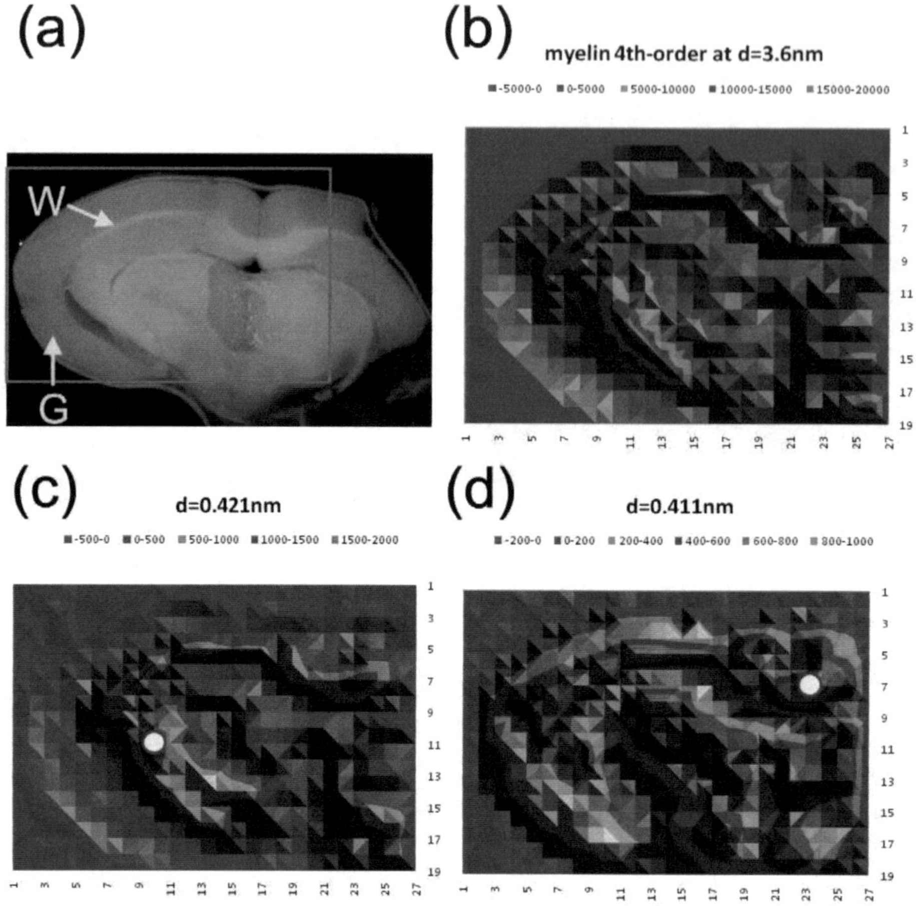

Figure 4. An example of scanning results of a coronal section of rat cerebrum. (a) A micrograph of the specimen. The area boxed by the red line was scanned. W denotes white matter, G gray matter. (b) The distribution of intensity of the 4th-order reflection from myelin (at 3.6 nm). The sample was scanned from the top left to the right bottom, with 27 steps horizontally and 19 steps vertically. The size of each step was 0.5 mm. (c) The distribution of intensity of the lipid peak at d=0.421 nm. (d) Distribution of the intensity of the lipid peak at d=0.411 nm. The yellow dots in (c) and (d) are where the diffraction patterns of myelin and gray matter in Figure 3 were obtained.

4. Discussion

It is not clear why only formalin-fixed brain gives sharp rings due to the lateral packing of lipids. Since brain lipid bilayers have a considerable amount of protein, fixation of proteins, especially cross-linking of side chains of lysine and arginine, is conceivable. On the other hand, as there has been reports that formaldehyde combines with unsaturated lipids at the double bond [12,13], effects on lipid molecules are also possible. Since there has not been a report showing sharp lipid peaks before, the extended fixation used in this study may be effective in inducing lateral orders of lipid molecules..

There is a marked difference in the lipid composition between gray and white matters. In gray matter, glycerophosphatides are dominant with some sphingolipids and cholesterol (ratio 21:5:7 in dry weight percent in adult human brain [14]), while in white matter there are much more sphingolipids and cholesterol with the ratio of 21:21:15. This is mainly due to the composition in myelin which has a ratio of 24:24:19.

It has been found that the presence of cholesterol in DPPC (dipalmitoylphosphatidylcholine) / DOPC (dioleoylphosphatidylcholine) model membrane system causes disorder in chain-packing and increases the lateral distance between lipid chains [15]. It was reported that with DPPC:DOPC=1:1 the X-ray peak due to lateral packing was at d=0.415 nm while with DPPC:DOPC:cholesterol=1:1:0.5 (molar ratio), the peak was at 0.428 nm. Thus, the higher cholesterol content may be the reason the 0.421nm peak is stronger than the 0.411nm peak in myelin in the present study.

This is the first study to investigate lipid structures in different regions of brain and preliminary in nature. It is possible that only a small fraction of lipids in these regions is contributing to the sharp peaks. Presence of other reflections show that there are other types of components in brain. Since brain is a very complicated organ, there must be different lipids playing different roles. Especially in gray matter, lipids may be not only in cellular membranes that transmit electrical signals but also in subcellular components. Information on the structure and composition of lipids in the subcellular organs is scarce. However, the structural difference between the lipids in gray and white matters shows that there are fundamental variations which may be related to the functions of lipids in these regions.

Acknowledgements
I thank Dr. Takaaki Hikima and Dr. Hiroyasu Masunaga for help at BL45XU.

References

[1] Bras W, Dolbnya I P, Detollenaere D, van Tol R, Malfois M, Greaves G N, Ryan A J, and Heeley E 2003 *J. Appl. Crystallogr.* **36** 791
[2] Yagi N, Ohta N, Matsuo T, Tanaka T, Terada Y, Kamasaka H, To-o K, Kometani T, and Kuriki T 2009 *J. Synchrotron Radiat.* **16** 398
[3] Campanella R 1992 *J. Neurosurg. Sci.* **36** 11
[4] Chia L S, Thompson J E, and Moscarello M A 1984 *Proc. Natl. Acad. Sci. USA* **81** 1871
[5] De Felici M, Felici R, Ferrero C, Tartari A, Gambaccini M, and Finet S 2008 *Phys. Med. Biol.* **53** 5675
[6] Falzon G, Pearson S, Murison R, Hall C, Siu K, Round A, Schültke E, Kaye A H, and Lewis R 2007 *Phys. Med. Biol.* **52** 6543
[7] Siu K K W, Butler S M, Beveridge T, Gillam J E, Hall C J, Kaye A H, Lewis R A, Mannan K, McLoughlin G, Pearson S, Round A, R, Schültke E, Webb G I, and Wilkinson S J 2005 *Nucl. Instrum. Meth. A* **548** 140
[8] Kumasaka T, Yamamoto M, Yamashita E, Moriyama H, and Ueki T 2002 *Structure* **10** 1205
[9] Yamamoto M, Kumasaka T, Yamazaki H, Sasaki K, Yokozawa Y, and Ishikawa T 2001 *Nucl. Instrum. Meth. A* **467-468** 1160
[10] Caspar D L D and Kirschner d A 1971 *Nature New Biol.* **231** 46
[11] Inouye H and Kirschner d A 1984 *J. Neurocytol.* **13** 883
[12] Jones D 1972 *Histochem. J.* **4** 421
[13] Wolman M and Greco J 1952 *Biotech. Histochem.* **27** 317
[14] O'Brien J S and Sampson E L 1965 *J. Lipid Res.* **6** 537
[15] Chen L, Yu Z, and Quinn P J 2007 *Biochim. Biophys. Acta* **1768** 2873

Journal of Physics: Conference Series **272** (2011) 012010 doi:10.1088/1742-6596/272/1/012010

Chain dimensions in free and immobilized brush states of polysulfobetaine in aqueous solution at various salt concentrations

Y Terayama[1], H Arita[1], T. Ishikawa[1], M Kikuchi[2], K Mitamura[2], M Kobayashi[2], N L Yamada[3] and A Takahara*[1,2,4]

[1] Graduate School of Engineering, Kyushu University, 744 Motooka, Nishi-ku, Fukuoka 819-0385, Japan

[2] JST, ERATO Takahara Soft Interfaces Project, 744 Motooka, Nishi-ku, Fukuoka 819-0385, Japan

[3] Neutron Science Laboratory, High Energy Accelerator Research Organization, 1-1 Oho, Tsukuba, Ibaraki 305-0801, Japan

[4] Institute for Materials Chemistry and Engineering, Kyushu University, 744 Motooka, Nishi-ku, Fukuoka 819-0385, Japan

E-mail: takahara@cstf.kyushu-u.ac.jp

Abstract. The chain dimensions of free and immobilized polysulfobetaine in aqueous solution at various salt concentrations were investigated by size-exclusion chromatography with multiangle light scattering and neutron reflectivity measurement, respectively. The dependence of the *z*-average mean square radius of gyration ($<S^2>_z^{1/2}$) on the weight-average molecular weight (M_w) of free poly(3-(*N*-2-methacryloyloxyethyl-*N*,*N*-dimethyl)ammonatopropanesulfonate) (MAPS) in aqueous solution at salt concentrations of 74, 100, 200, and 500 mM was described by the perturbed wormlike chain model using the chain stiffness parameter λ^{-1}, the molar mass per unit contour length M_L, and the excluded volume effect *B*. *B* increased from 0 to 1.8 nm with increasing salt concentration to 500 mM due to the screening of attractive electrostatic interaction between ammonium cations and sulfonyl anions by salt ions. The swollen structure of the poly(MAPS) brush in D_2O changed from a shrunken state to a relatively extended state with increasing salt concentration from 0 to 500 mM NaCl/D_2O solution. The thickness of the swollen poly(MAPS) brush in 500 mM NaCl/D_2O was 9.0 times greater than $2<S^2>_z^{1/2}$ of free poly(MAPS) due to high osmotic pressure generated by the excluded volume effect of densely grafted polymer chains.

1. Introduction

Aqueous solution of poly(3-(*N*-2-methacryloyloxyethyl-*N*,*N*-dimethyl)ammonatopropanesulfonate) (MAPS) exhibits a unique salt dependency compared with the typical polyelectrolyte. In deionized water, the poly(MAPS) chain is insoluble due to the strong attractive inter- and intramolecular electrostatic interaction between ammonium cations and sulfonyl anions. With increasing salt concentration in aqueous solution, especially above theta concentration, poly(MAPS) shows enhanced solubility in aqueous solution and greater chain dimensions because the attractive electrostatic interaction among sulfobetaine groups is neutralized by the screening effect of salt ions [1]. Kato et al.

estimated the mean square radius of gyration $\langle S^2\rangle_z^{1/2}$ and theta concentration of poly(MAPS) to be 60 mM NaCl aqueous solution using static light scattering for polymers ranging from 5×10^4 to 3×10^5 in weight-average molecular weight (M_w); however, the molecular weight distribution (MWD) of the polymers was not clarified in their report. Polymers with broad MWD are unfavorable for the estimation of chain dimensions. In the present study, we precisely estimated the theta concentration and mean square radius of gyration $\langle S^2\rangle_z^{1/2}$ of poly(MAPS) in NaCl aqueous solution using size-exclusion chromatography combined with multiangle light scattering (SEC-MALS) for polymers with a wide range of molecular weight ($M_w = 1\times10^4 - 3\times10^6$) and narrow MWD ($M_w/M_n < 1.2$).

On the other hand, when poly(MAPS) is covalently tethered to a solid surface to form a brush, the thickness of the brush layer in the solution is affected by the solvent quality and ionic strength [2]. The poly(MAPS) brush chains in aqueous salt solution would be considerably extended in the perpendicular direction against the substrate over the radius of gyration of the corresponding free polymer due to the reduction in attractive intra- and intermolecular interaction and excluded volume effect. In this study, the effect of salt concentration on the swollen thickness of the poly(MAPS) brush with relatively narrow MWD was estimated by neutron reflectivity (NR) measurement and compared with the dimensions of unbound poly(MAPS). The depth profiles of the swollen brush structures were also examined from the scattering length density curves.

2. Experimental

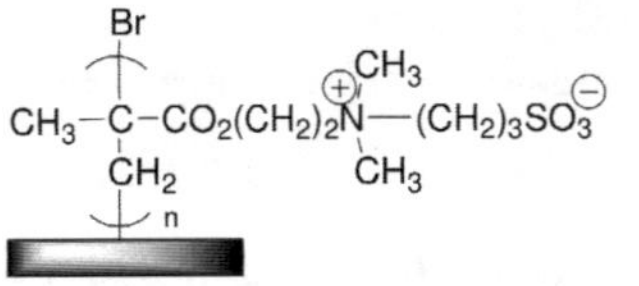

Scheme 1. Poly(MAPS) brush.

Surface-initiated atom transfer radical polymerization (ATRP) of MAPS [3] from (2-bromo-2-methyl)propionyloxyhexyltrimethoxy-silane-immobilized silicon wafer and quartz plate was carried out using CuBr and 2,2-dipyridyl in 1-hexyl-3-methylimidazolium chloride / 2,2,2-trifluoroethanol (TFE) (1/9, wt/wt) mixture at 60°C for 12 h under dry argon atmosphere to give a poly(MAPS) brush with M_w of 2.69×10^5, relatively narrow MWD ($M_w/M_n = 1.23$), and thickness of 41 nm as determined by ellipsometry in air.

Free poly(MAPS) was prepared by means of free radical polymerization in TFE using 2,2'-azobis(isobutyronitrile) as an initiator and was fractionated to 10 parts with different molecular size by a recycling-preparative HPLC system (Japan Analytical Industry Co., Ltd., LC-9104) equipped with an RI detector (JAI RI-7s) using two columns (JAIGEL GS620-50L, GS320-50L) in 200 mM aqueous NaCl solution as an eluent at a flow rate of 20 mL min^{-1} at room temperature.

SEC was carried out using an RI detector (Shimadzu, RID-10A) equipped with three columns (Tosoh, TSKgel G3000PW$_{XL}$, G5000PW$_{XL}$, and G6000PW$_{XL}$) in aqueous NaCl solutions as the eluent at a flow rate of 0.8 mL min^{-1} at 40°C. M_w and z-average mean square radius of gyration ($\langle S^2\rangle_z^{1/2}$) of the fractionated poly(MAPS) were determined by SEC-MALS (Wyatt Technology, DAWN-EOS, wavelength: λ = 690 nm) at 25°C. The Rayleigh ratio $R(90)$ at a scattering angle of 90° was determined using pure toluene. The sensitivity and dead volume of 16 detectors were corrected using the scattering intensity of 0.30 wt% NaCl aqueous solution of a poly(2-(*N*-2-methacryloyloxyethyl-*N*,*N*-dimethyl)ammonatoethanesulfonate) standard with $M_w = 4.82\times10^4$ g mol^{-1} and $M_w/M_n = 1.10$. Polymer solutions with mass concentration of about 1×10^{-3} g cm^{-3} were injected into SEC columns using a sample loop of 100 μL. The specific refractive index increment (*dn*/*dc*) of the poly(MAPS) ($M_w = 1.36\times10^5$ g mol^{-1}, $M_w/M_n = 3.07$) in 74, 100, 200, and 500 mM NaCl aqueous solution at 25°C measured by a differential refractometer (Otsuka Electronics DRM-1020, wavelength: λ = 633 nm) was 0.121_4, 0.121_2, 0.120_3, and 0.117_6 cm^3 g^{-1}, respectively. These *dn*/*dc* values were used for the analysis of SEC-MALS even though the laser wavelength was different from that of DRM, because the difference between *dn*/*dc* values at λ = 633 nm and λ = 690 nm is negligible.

NR measurement was carried out using the ARISA-II time-of-flight type reflectometer at BL16 in MLF/J-PARC providing 25-Hz-pulsed neutron radiation at ~120 kW. The wavelength of the incident neutrons was tuned at 0.2–0.88 nm by a disk chopper. A neutron beam was irradiated from the quartz to the interface between D_2O and poly(MAPS) brush on quartz glass and the reflected neutrons were collected by a 2D scintillation detector. The incident slits were adjusted to maintain a 50-mm footprint size on the sample surface at any incident angle (typically 0.30, 0.75, 1.20°). The NR profiles were analyzed by fitting the calculated reflectivity from the model scattering length density (SLD) profiles to the data, using Parratt32 software.

3. Results and discussion

3.1. Chain dimensions of free poly(MAPS) in aqueous solution at various salt concentrations

The chain dimensions of free poly(MAPS) in 74, 100, 200, and 500 mM NaCl aqueous solutions were investigated by SEC-MALS measurement using 10 samples with various M_w ranging from 1.11×10^4 to 5.45×10^6 and with narrow MWD (M_w/M_n < 1.2) prepared by fractionation using a recycling-preparative HPLC system. Figure 1(a) shows typical SEC curves of poly(MAPS) and the plot of M_w versus retention volume (*RV*) in 74 and 500 mM NaCl aqueous solutions calculated from the SEC curves of 10 samples with different M_w. Despite the same M_w, the SEC curve of poly(MAPS) in 500 mM NaCl aqueous solution appeared at a lower RV region compared with the 74 mM solution. The calibration curve for the 500 mM NaCl aqueous solution clearly shifted to a lower *RV* region compared to that for the 74 mM solution, indicating that the poly(MAPS) chains in the 500 mM NaCl aqueous solution had expanded.

The $\langle S^2\rangle_z^{1/2}$ of the poly(MAPS) in 74, 100, 200, and 500 mM NaCl aqueous solution was estimated using equation (1):

$$P(q)^{-1/2} = 1 + \frac{1}{6}\langle S^2\rangle_z q^2 \text{ and } q = \frac{4\pi n_0 \sin(\theta/2)}{\lambda} \quad (1)$$

where q, $P(q)$, n_0, and θ are the scattering vector, single particle scattering function, refractive index of solvent, and scattering angle of detector, respectively. Figure 1(b) shows typical plots of $P(q)^{-1/2}$ vs. q^2 for the poly(MAPS) at different *RV* from 18.9 to 22.3 mL when 74 mM NaCl aqueous solution was used as an eluent. The $\langle S^2\rangle_z^{1/2}$ was determined by the initial slope between $P(q)^{-1/2}$ vs. q^2.

Figure 2(a) shows the dependency of $\langle S^2\rangle_z^{1/2}$ on M_w of poly(MAPS) in 74, 100, 200, and 500 mM NaCl aqueous solution at 25°C. In general, $\langle S^2\rangle_z^{1/2}$ is related to M_w^{α}, where the scaling exponent α varies with solvent quality; α = 0.5 indicates theta conditions. We found that 74 mM NaCl aqueous solution revealed the scaling exponent to be 0.5 from the slope of logarithmic plots of $\langle S^2\rangle_z^{1/2}$ and M_w, as shown in figure 2(a). Therefore, we established 74 mM NaCl aqueous solution as the theta concentration of poly(MAPS). Kato et al. previously determined the theta concentration for poly(MAPS) by static light scattering at 30°C to be 60 mM of NaCl aqueous solution [1]. The slight difference between 74 mM in our study and 60 mM in theirs might be attributed to the broader MWD of poly(MAPS) compared with our experiment.

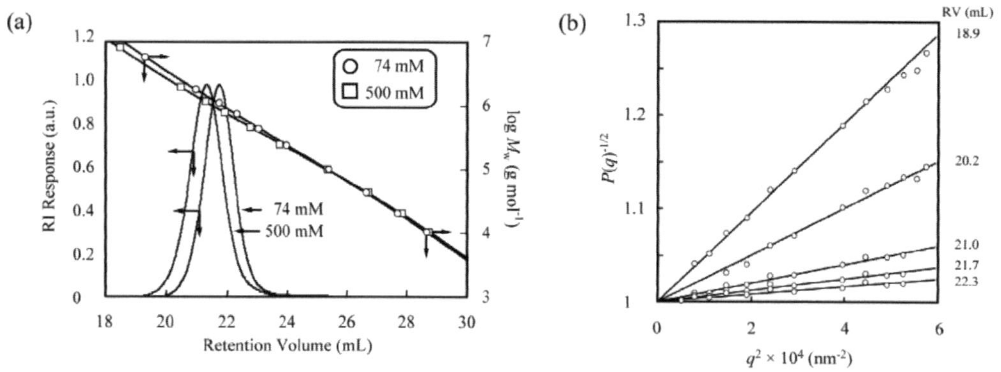

Figure 1. (a) SEC curves of fractionated poly(MAPS) ($M_w = 1.21\times10^6$, $M_w/M_n = 1.17$) in 74 and 500 mM aqueous NaCl solution, together with the absolute calibration curves of the poly(MAPS) in aqueous NaCl solution with 74 (circles) and 500 mM (squares) determined by SEC-MALS. (b) Angular dependence of $P(q)^{1/2}$ for the indicated *RV* of the poly(MAPS) in 74 mM aqueous NaCl solution at 25°C by SEC-MALS.

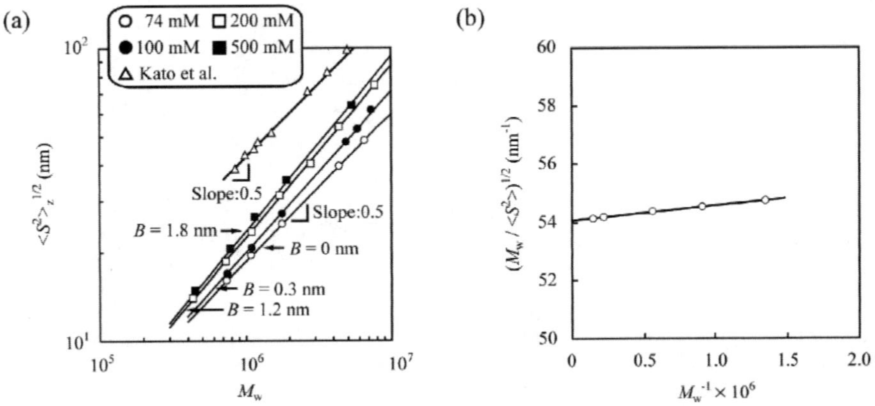

Figure 2. (a) M_w dependence of $<S^2>_z^{1/2}$ for the poly(MAPS) in 74, 100, 200, and 500 mM aqueous NaCl solution at 25°C as experimental data, and in 60 mM at 30°C as reported by Kato et al. [2]. The solid curves are the theoretical values for the perturbed wormlike chain model as calculated by equation (2). (b) Murakami Plot of $[M_w/<S^2>_z]^{1/2}$ vs. M_w^{-1} for the poly(MAPS) in 74 mM aqueous NaCl solution at 25°C.

For the chain dimensions of the polymer under theta conditions, the mean square radius of gyration $<S^2>_{KP}$ can be described by the unperturbed wormlike chain (Kratky-Porod chain) model followed by equation (2) [4]:

$$\langle S^2 \rangle_{KP} = \frac{L}{6\lambda} - \frac{1}{4\lambda^2} + \frac{1}{4\lambda^3 L} - \frac{1}{8\lambda^3 L^2}\left[1 - \exp(-2\lambda L)\right] \tag{2}$$

where λ^{-1} and L are the chain stiffness parameter (Kuhn segment length) and the contour length, respectively. When λL is greater than 2, the maximum error between the theoretical and exact values is less than 1%. Therefore, equation (2) approximates equation (3) (Murakami plot) [5]:

$$\left[\frac{M}{\langle S^2 \rangle}\right]^{1/2} = (6\lambda M_L)^{1/2}\left(1 + \frac{3M_L}{4\lambda L M}\right) \tag{3}$$

where M and M_L are the molecular weight of polymer and the molar mass per unit contour length, respectively, and are defined as $L = M/M_L$. The λ^{-1} and M_L can be evaluated from the slope and intercept of the plot of $[M/<S^2>_z]^{1/2}$ vs. M^{-1} using equation (3). Figure 2(b) shows the plot of $[M_w/<S^2>_z]^{1/2}$ vs. M_w^{-1} for the poly(MAPS) in 74 mM NaCl aqueous solution. These plots were well aligned on the theoretical line described by equation (3) using 5.1 nm of λ^{-1} and 2470 g mol^{-1} of M_L corresponding to the experimental $M_w/<S^2>_z$. The estimated λ^{-1} = 5.1 nm of poly(MAPS) in 74 mM NaCl aqueous solution was larger than the λ^{-1} = 1.5 nm of PMMA under theta conditions [6]. The contour length per MAPS monomer is 0.11 nm, which is smaller than the 0.125 nm of the PMMA chain with the all-*trans* conformation. These results indicated that the experimental M_w dependence of $<S^2>_z^{1/2}$ under theta conditions was quantitatively described by the unperturbed wormlike chain model.

The M_w dependence of $<S^2>_z^{1/2}$ in 100, 200, and 500 mM should be described by the unperturbed wormlike chain model modified by the quasi-two-parameter (QTP) theory using the excluded volume strength (B) [7-10]. Figure 2(a) shows the experimental plots of $<S^2>_z^{1/2}$ in 100, 200, 500 mM and the corresponding theoretical straight line calculated using an unperturbed wormlike chain model based on the QTP theory. All $<S^2>_z^{1/2}$ were well represented by theoretical lines when B in 100, 200, and 500 mM was 0.3, 1.2, and 1.8 nm, respectively. The increase in B parameter indicated that the intramolecular excluded volume effect increased with increasing salt concentration from 100 to 500 mM due to the screening effect of salt ions in the aqueous solution. The experimental M_w dependence of $<S^2>_z^{1/2}$ for the poly(MAPS) in NaCl aqueous solution at all salt concentrations was quantitatively described in terms of the perturbed wormlike chain model using the three parameters of λ^{-1}, M_L, and B.

3.2. Swollen state of poly(MAPS) brush on the substrate at aqueous solution interface

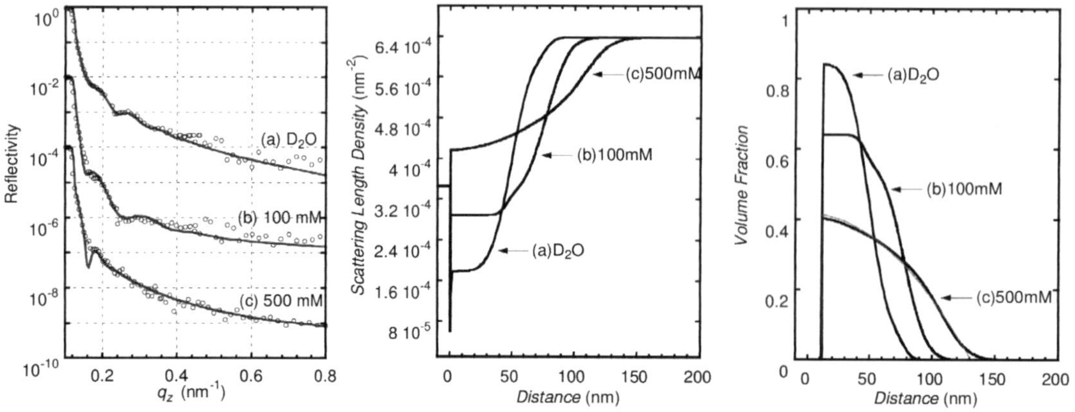

Figure 3. NR curves (left) of poly(MAPS) brush in (a) D_2O, (b) 100 mM NaCl/D_2O, (c) 500 mM NaCl/D_2O. Their corresponding neutron SLD profiles (center) and volume fraction profiles of MAPS (right) along with the distance from the quartz surface. The fitting line of the volume fraction profile in 500 mM poly(MAPS) was calculated using the parabolic function $\varphi = \varphi_0(1-(z/h)^2)$, where φ_0 is the distance from the substrate $z = 0$, and h is the cut-off thickness of the swollen brush layer.

Figure 3 shows the NR curves of the poly(MAPS) brush in D_2O, 100 mM, and 500 mM NaCl/D_2O and their corresponding fitting curves calculated by assumed SLD profiles consisting of a quartz layer (3.40×10^{-4} nm^{-2}), initiator layer (0.60×10^{-4} nm^{-2}), gradient brush layer, and D_2O layer (6.38×10^{-4} nm^{-2}). The SLD in D_2O increased drastically from 1.70×10^{-4} to 6.38×10^{-4} nm^{-2} at 85 nm. With increasing salt concentration to 100 mM, the SLD profile gradually increased from 3.05×10^{-4} to 6.38×10^{-4} nm^{-2} at 120 nm. With a further increase in salt concentration to 500 mM, the SLD reached 150 nm with increasing SLD from 4.0×10^{-4} to 6.38×10^{-4} nm^{-2}. Swollen brush thickness h was estimated at 130 nm by volume fraction profile using parabolic function fitting. In D_2O, the brush chains interact with neighboring polymer chains due to the attractive electrostatic interaction of sulfobetaine groups to form shrunken structures. The brush chain length in 100 mM NaCl/D_2O was slightly increased in the direction of the D_2O interface due to screening of the electrostatic interaction by salt ions, while inside the brush chain a shrunken state was maintained. On the other hand, the brush chain was further extended at 150 nm in 500 mM NaCl/D_2O. The volume fraction profile of MAPS in 500 mM NaCl/D_2O shows a parabolic function curve indicating that the brush chain was completely swollen in the aqueous solution [11].

Free poly(MAPS) dissolved in the NaCl aqueous solution above 74 mM NaCl aqueous solution; however, the near substrate region of poly(MAPS) brush chains in 100 mM NaCl/D_2O seemed to remain in a shrunken state. We supposed that densely-grafted poly(MAPS) chains strongly interacted with each other to generate strong attractive electrostatic interaction among sulfobetaine groups. On the other hand, the poly(MAPS) brush in 500 mM NaCl/D_2O was swollen because concentrated salt ions effectively screened the electrostatic interaction inside the brush chains. The region of the brush chains reached 131 nm from the substrate, which was 9.0 times greater compared with 14.6 nm of $2\langle S^2\rangle_z^{1/2}$ of free poly(MAPS) corresponding to the same M_w of the estimated poly(MAPS) brush in 500 mM NaCl aqueous solution. The graft density of the poly(MAPS) brush σ was 0.26 chains nm^{-2} calculated by $\sigma = d\times N_A\times L\times10^{-21}/M_n$ where d, N_A, and L are the assumed bulk density, Avogadro's number, and the dry thickness of the brush, respectively, i.e., the occupied area of a single brush chain was 3.84 nm^2, which is smaller than the chain dimensions of the free poly(MAPS) in 500 mM NaCl aqueous solution. We supposed that high osmotic pressure among the densely grafted polymer chains resulted in the relatively extended structure in aqueous solution.

Salt concentration dependency of chain dimensions in aqueous solution varies according to the chemical structure of ionic groups of polyelectrolytes. The uniqueness of poly(MAPS) can be identified by comparing the salt concentration dependency with that of other polyelectrolyte chain conformations. For example, poly(2-methacryloyloxyethyl phosphorylcholine) (MPC) brush, which is a zwitter-type polyelectrolyte as well as poly(MAPS), in salt aqueous solution maintains a relatively extended chain structure over a wide range of ionic strength in the solution probably due to weak electrostatic static interaction among the phosphobetaine groups [12, 13]. In the case of a typical anionic polymer, poly(styrenesulfonate sodium salt) (SSNa), poly(SSNa) brush in deionized water forms a stretched structure due to the strong repulsion between sulfonate monomers [14], while the brush takes on a more coiled conformation, similar to electronically neutralized chains, in 5.0 M NaCl aqueous solution by screening of electrostatic interchain repulsion. In contrast, the poly(MAPS) brush in aqueous media transform from collapsed to extended conformation with increase in salt concentration due to the reduction in attractive electrostatic interaction among sulfobetaine groups by hydrated salt ions. This unique solution behavior could be applicable to friction or adhesion control of the surface.

4. Conclusion

The thickness of the swollen poly(MAPS) brush estimated by NR measurement increased at increasing salt concentration from 0 to 500 mM. However, the brush chains near the substrate in 100 mM NaCl aqueous solution remained in a shrunken state due to the strong attractive electrostatic interaction among betaine groups, even though the 100 mM NaCl aqueous solution would be a good solvent for free poly(MAPS) because the theta concentration was 74 mM. This attractive electrostatic interaction was further screened in 500 mM NaCl/D_2O solution to give an electrically neutral polymer brush. The swollen thickness of the poly(MAPS) brush achieved a value 9.0 times greater than $2\langle S^2\rangle_z^{1/2}$ of free poly(MAPS), indicating that the brush chains in 500 mM NaCl/D_2O extended in the perpendicular direction from the substrate due to high osmotic pressure generated by densely grafted polymer chains.

Acknowledgements

The present work is supported by a Grant-in-Aid for the Global COE Program, "Science for Future Molecular Systems" from the Ministry of Education, Culture, Science, Sports and Technology of Japan. YT acknowledges support from the Japan Society for the Promotion of Science. The experiments at J-PARC were performed under the approval of the Neutron Scattering Program Advisory Committee of IMSS, KEK (Proposal No. 2009S08).

References

[1] Kato T and Takahashi A 1996 *Ber. Bunsenges. Phys. Chem.* **100** 784
[2] Terayama Y, Kikuchi M, Kobayashi M, Hino M and Takahara A 2009 *J. Phys: Conf. Ser.* **184** 012011
[3] Terayama Y, Kikuchi M, Motoyasu K and Takahara A, 2011 *Macromolecules*, **44**,104
[4] Benoit H and Doty P 1953 *J. Phys. Chem.* **57** 958
[5] Murakami H, Norisuye T and Fujita H 1980 *Macromolecules* **13** 345
[6] Tamai Y, Konishi T, Einaga Y, Fujii M and Yamakawa H 1990 *Macromolecules* **23** 4067
[7] Yamakawa H and Stockmayer W H 1972 *J. Chem. Phys.* **57** 2843
[8] Domb C and Barrett A J 1976 *Polymer* **17** 179
[9] Yamakawa H and Shimada J 1985 *J. Chem. Phys.* **83** 2607
[10] Shimada J and Yamakawa H 1986 *J. Chem. Phys.* **85** 591
[11] Milner S T, Witten T A and Cate M E 1988 *Macromolecules* **21** 2610
[12] Kobayashi M, Terayama Y, Hino M, Ishihara K and Takahara A 2009 *J. Phys: Conf. Ser.* **184** 012010
[13] Matsuda Y, Kobayashi M, Annaka M, Ishihara K and Takahara A, 2008 *Langmuir*, **24**, 8772
[14] Tran Y, Auroy P, and Lee L-T 1999 *Macromolecules* **32** 8952

Journal of Physics: Conference Series **272** (2011) 012011 doi:10.1088/1742-6596/272/1/012011

Transformation from Multilamellar to Unilamellar Vesicles by Addition of a Cationic Lipid to PEGylated Liposomes Explored with Synchrotron Small Angle X-ray Scattering

Mina Sakuragi[1], Kazunori Koiwai[2], Kouji Nakamura[2], Hiroyasu Masunaga[3], Hiroki Ogawa[3] and Kazuo Sakurai[1*]
[1]The University of Kitakyushu, [2]Terumo Corporation R & D Center, [3]JASRI/SPring-8

E-mail: sakurai@env.kitakyu-u.ac.jp

Abstract. PEGylated liposomes composed of a benzamidine derivative (TRX), hydrogenated soybean phosphatidylcholine (HSPC), and N-(monomethoxy-polyethyleneglycolcarbamyl) distearoyl phosphatidylethanolamine (PEG-PE) were examined in terms of how the addition of TRX affects their structures with small angle x-ray scattering (SAXS) as well as transmission electron microscopy (TEM). TEM images showed the presence of unilamella vesicles for both with and without TRX, though a small amount of multilamella vesicles were observed in absence of TRX. We analyzed SAXS profiles at contained TRX composition combined with contrast variation technique by adding PEG solution and unilamella vesicle model could be reproduced. Subsequently, we analyzed SAXS profiles at no TRX composition. The mixture model of unilamella and multilamella vesicle was reconstructed and we estimated about 10 % multilamella vesicles from a fitting parameter.

1. Introduction

Phospholipids have a distinctive propensity to form bilayer membranes in aqueous solutions [1]. This molecular character has been used to prepare lipid vesicles, or liposomes; defined as aqueous compartments enclosed by the bilayer membrane. Liposomes can be formed by sonicating lipid-dispersed solutions to give closed vesicles that are in most case quite uniform in size. The liposomes formed by this method are generally unilamellae and spherical in shape with a diameter of about 50 - 200 nm, denoted by ULV (unilamellar vesicle) [2]. In some cases, concentric multilamellar vesicles (MLV) with an onion-like layered membrane or a large unilamellar vesicle contains several smaller oligo-vesicles without concentric arrangement (OVV) can be formed [3]. On strict definition, only ULM should be called liposomes; however, we treat MLS and OVV as degenerated liposomes.

Liposomes have been used as a drug carrier because of capability to encapsulate hydrophobic compounds within their bilayers [4] as well as hydrophilic compounds within their internal aqueous areas [5]. Liposomes are only a protocol to use *in vivo* gene delivery system [6-8]. Recently, to prolong their blood circulation time, poly(ethylene glycol) modification (PEGylation) on the liposome surface is commonly applied [9]. PEGylation causes the liposome surface less susceptible to immunological attack and thus to reduce clearance by reticuloendothelial system [10]. The molecular mechanism of such effects is considered owing to steric hindrance or the excluded volume effect of the hydrated PEG chains on the surface [11]. In addition, PEG has other advantages such as variability in chain length and abundance of protocols for its chemical derivatization [12].

It was reported that benzamidine and benzyle-amine derivatives (the former is denoted by TRX) can be used as a transfection regent with a better efficiency and less toxic than commercial products [13-16]. Among others, Kawahara et al [17] examined pharmacokinetics of PEGylated TRX/HSPC liposome and found that the addition of TRX to PEGylated liposomes induces drastic change in the half-life in the blood circulation. Here, HSPC is hydrogenated soybean phosphatidylcholine that is well-known to form liposomes [18]. Although this change may be due to the cationic nature of added TRX, it is important to examine how addition of TRX changes the liposome structures. This paper examines PEGylated TRX/HSPC liposome in terms of how the addition of TRX affects its structures by using small angle x-ray scattering (SAXS) as well as transmission electron microscopy (TEM).

2.Material and Methods

Material: TRX having two pentadecyloxy tails attaching at meta-position was synthesized with a reported method [19]. N-(monomethoxy- polyethyleneglycolcarbamyl) distearoyl phosphatidylethanolamine (PEG-PE) with a purity of more than 95% was obtained from Genzyme Company (Switzerland) and HSPC was purchased from Lipoid KG (the Netherlands). The chemical structures of these lipids were shown in Figure 1. The PEGylated liposome composed of TRX/HSPC (and of only HSPC for comparison) were prepared with the reported method described in elsewhere [17]. Here, non-PEGylated one was first prepared and secondary filtrated with an extrusion process (Extruder T-10; Lipex Biomembranes, Canada) using a 0.2 μm membrane filters (Whatman, U.S.A.) and PEG-PE was then added at 1.50 mol% and unencapsulated PEG-PE was removed by gel filtration. The compositions of liposomes and code names are shown in Table.1. Phosphate buffered saline (PBS) tablet was purchased from Sigma-Aldrich, Japan and 1/10 PBS at pH = 7.4 was used as a solvent. Homo-polyethylenegycol with the weight average molecular weight of 400 (PEG400) was purchased from Wako Pure Chemical Industries, Ltd.

Characterization of Liposomes: The particle size of PEG-liposome was determined dynamic light scattering (DLS) with a Zetasizer Nano Series (SYSMEX CORPORATION). In this instrument, the scattering angle is fixed at 173° and the obtained time correlation data was analyzed with the cumulant method by use of the attached program to obtain the size distribution profile. Transmission electron microscopy (TEM) was carried out for the specimens made with the following procedure; PEG-liposome solutions were diluted to 0.4 mM with 1/10 PBS then added with an equal volume of 2 % $(NH_4)_2MoO_4$ solution and left for 1 h to negatively stain the samples.

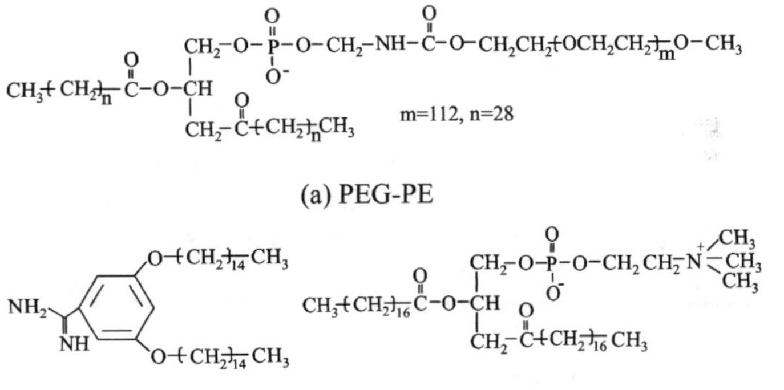

Figure 1. Chemical structures of lipids used in this study.

Table 1. The compositions of liposomes and their code names.

sample name	sample	TRX / HSPC (wt%)
0-TRX	HSPC / PEG-PE	0
12-TRX	HSPC / TRX / PEG-PE	11.6
19-TRX	HSPC / TRX / PEG-PE	18.7

The obtained solution was dropped onto a carbon-coated copper TEM grid and then placed onto the surface of a copper grid for 15 min and the excess solution was drawn off using a filter paper. After freeze drying, TEM images were observed with a JEOL JEM 3010 at an accelerating voltage of 200 kV.

Small angle x-ray scattering: SAXS measurements were carried out at 40B2 SPring-8 with 0.7 or 4 m camera using a Rigaku imaging plate (30×30 cm, 3000×3000 pixels) as a detector. The wavelength of the beam was 1.0 or 0.71 Å, the exposure time was 120 sec or 300 sec, and the relative X-ray intensity and the sample transmittance were determined with two ion-chambers located in front of and behind the sample. A bespoke chamber for SAXS measurements was constructed that enables a sample to be introduced in vacuum and scattering measured from it [20]. A quartz capillary cell (diameter of 2 mm, Hilgenberg GmbH) was used for all measurements.
Fitting Models and Electron Density Calculation: The form factor for multilayer spherical model (*Pc*) is given by Eq 1[21] [22] as a summation of concentric multiple spheres with number of *k*.

$$P_s(q) \propto \left[\sum_{i=1}^{k} \frac{(\rho_i - \rho_{i+1}) V_i j_1(qR_i)}{qR_i} \right]^2 \qquad (1)$$

Here, j_1 is the first order spherical Bessel function. ρ_i, V_i, and R_i are the electron density, the volume, the radius of the *i*-th sphere, respectively. ρ_{k+1} is the electron density of solvents (i.e., $\rho_{k+1} = \rho_{\mathrm{sol}}$) this equation, the width of the *i*-th layer (t_i) is given by $t_i = R_i - R_{i-1}$ except for $i = 1$, since the first layer is essentially the inner compartment of the liposome. When applied to the data, we assume the Gaussian distribution only for R_1. One serious problem to use the multilayer model is that there are too many adjustable parameters to obtain a unique solution. Therefore, in order to reduce mathematical degree of freedom in Eq 1, we need rules and relation to correlate the parameters to real molecules.
The solvent electron densities of PBS and after PEG 400 added were calculated from the following equation;

$$\rho = d \sum_i \rho_{w,i} w_i \qquad (2)$$

Here, d (g/nm^3) is the density of the mixed solvent, and $\rho_{w,i}$ (e/g) and w_i are electron density per weight and the weight fraction for the *i* component, respectively. We applied a contrast variation technique to SAXS by adding PEG 400 to the solvent [23, 24]. PEG has a larger electron density than water (ρ_{water} = 334 and ρ_{PEG} = 370 e/nm^3), since they contain oxygen atoms which have a relatively larger scattering length than carbon and hydrogen atoms [25] .

3. Results

3-1 : Particle size analyses and electron microscopy

The obtained autocorrelation functions are plotted against the delay time in Figure 2a and the size-distributions determined are plotted in Figure 2b for 0-, 12-, and 19-TRX. Although three samples showed the similar narrow distribution curves around 100 nm, 0-TRX was slightly larger than the others. The peak-top sizes are 101, 87 and 84 nm, respectively. The narrow distribution is attributed to sonicating and filtrating (with a 0.2 μm membrane) treatments after the lipids were dispersed. We added PEG-PE after the filtration. This treatment did not change the size [17]. The small but appreciable difference between the presence and absence of TRX suggests that TRX can differentiate the liposome structures.

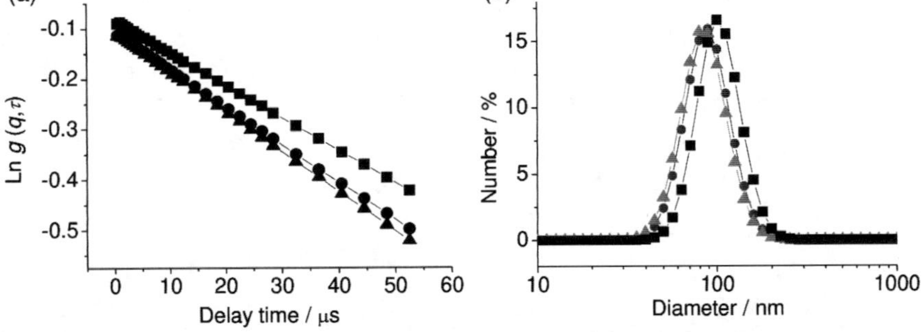

Figure 2. DLS from the three liposome solutions, ■;0-TRX, ●; 12-TRX, ▲; 19-TRX. (a) Autocorrelation function, $g(q,\tau)$ and (b) Particle size distributions.

Figure 3 compares the TEM images for the three liposomes. All of them showed a typical ring image for ULV liposome with a diameter around 100 nm. These sizes are consistent with the DLS results. When we carefully examined the images for 0-TRX, we found that a few of rings consisted of double line or their width broader than those of 12- and 19-TRX. The width of the ring for 12- and 19-TRX was about 7 nm, which is consistent with the molecular size of bilayers [26], while some of 0-TRX were about 13 nm as indicated in the Figure 3, almost twice of 12- and 19-TRX. This fact indicated that 0-TRX liposomes contain MLV. We sometimes clearly observed double ring images for 0-TRX although their population was seemed to be minor. However, based on TEM observation, it seemed difficult to obtain quantitative information.

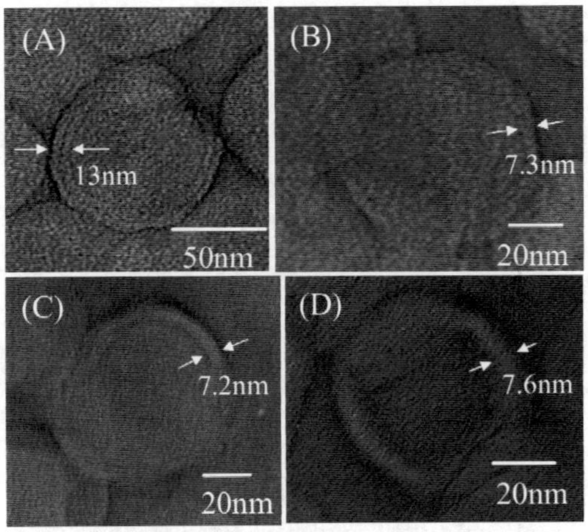

Figure 3. TEM images after staining with 1wt% ammonium molybdate for 0-TRX (A), (B), 12-TRX (C), and 19-TRX (D).

3-2 : Small angle X-ray scattering

Figure 4a compares the SAXS profiles for the three liposomes with the same concentration (10 mM) and Figure 4b and c show the concentration dependence of 0- and 12-TRX. The profiles of 12- and 19-TRX in Figure 4a show a typical scattering from ULV, i.e., the scattering from the outermost sphere at $q < 0.3$ nm^{-1} and the large undulations from bilayer membrane around $0.3 < q < 2$ nm^{-1} and smaller ones at $q > 2$ nm^{-1}. The Ginner region must be observed below the measuring range judging from the size of the liposome of 70-80 nm (the Ginner region for this size is $q < 0.04$ - 0.05 nm^{-1}). The large undulation is reflecting from the cross sectional structure of the liposome. The smaller ones at the larger q region evidence that the structure is almost uniquely determined by the composition and the fluctuation and distribution are relatively small.

0-TRX has the almost same feature with 12- and 19-TRX, except for a different shape of the large undulation. It seems that there is a shoulder around $q = 1.0$ nm^{-1}. This type of shoulder can be due to diffraction between liposomes due to infinite concentrations. Eguchi et al [27] extensively studied the concentration dependence of the SAXS from charged micelles and found that the similar shoulder appeared on the large undulation with increasing the concentration due to diffraction (i.e., disappeared at low concentration and the scattering merged into the form factor around C = 7.6 wt % [27]). Figure 4b shows that the shoulder did not disappeared even at 0.22 mM (0.02 wt %), at which concentration should be low enough to eliminate the inter-particle diffraction. This fact leads that the shoulder can be ascribed to the MLV structure observed with TEM.

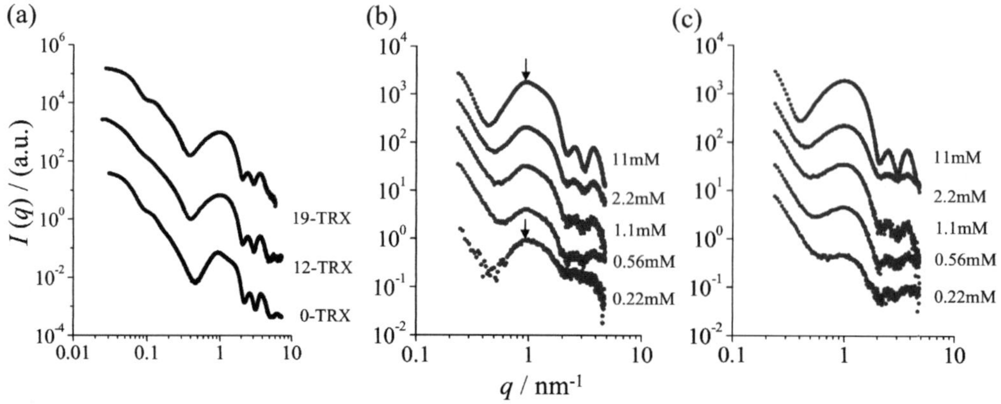

Figure 4. (a) Scattering profiles combined SAXS and USAXS. Comparison of the concentration dependence of the scattering profiles between 0-TRX (b) and 12-TRX (c). The arrows indicate the shoulder that was not observed in 12-TRX.

4. Discussions

4-1 : Contrast variation analysis of SAXS from TRX-12

A straightforward analysis might fit first for the 0-TRX scattering and then for 12-TRX by referencing the 0-TRX fitting parameters, because 0-TRX has less components than 12-TRX and thus might be simpler than 12-TRX. However, TEM showed presence of MLV in 0-TRX, meaning more fitting parameters, while 12-TRX presumably consists of only ULV, less fitting parameters. Therefore, we decided to analyze 12-TRX first.

As mentioned in the experimental section as well as depicted in Figure 5a, to describe ULV with a multilayer model, we need to set $k = 5$ in Eq 1 with having four geometric parameters; t_1: inner diameter of the liposome, t_2 and t_4 (assumed $t_2 = t_4$): the headgroup width, t_3: the alky chain domain width, four electron densities (ρ_1, ρ_2, ρ_3 and ρ_4), and ρ_{sol} : the solvent density (assumed $\rho_1 = \rho_{\mathrm{sol}}$), corresponding each layer. Additionally, since we can suppose the PEG block of PEG-PE forms a layer outside the outer head group, we introduced two parameters: the PEG layer width (t_5) and its electron density, (ρ_5) as presented in Figure 5a. Among these parameters, there are rules to follow in terms of molecular characters and Eq 2. The electron densities of bulk PEG and the PBS solvent are known as ρ_{PEG} = 370 e/nm^3 and ρ_{PBS} = 335 e/nm^3, respectively and thus $\rho_{sol} < \rho_5 < \rho_{PEG}$ because of the PEG dissolving in PBS. Furthermore, $\rho_{\mathrm{sol}} < \rho_2$ or ρ_4 and $\rho_3 < \rho_{\mathrm{sol}}$ should be hold owing to their constituent elements [23]. Although simple liposomes without PEG layers requires $\rho_2 = \rho_4$ due to symmetrical bilayer structures, this relation is not necessary maintained for PEGylated ones since the phosphoric group or hydrophobic tails are contained only in the 4-th layer.

Even with those rules to reduce freedom of the fitting parameters, there were too may adjustable parameters to obtain a unique combination of them. To solve this problem, we used a contrast variation technique by adding PEG to solvent. If the addition of PEG did not alter the liposome structures, we should be able to fit the all data with the same parameters except for ρ_{sol} that is determined by the solvent composition through Eq 2.

Figure 5b presents how the scattering profile was changed by addition of PEG 400. With adding PEG 400 to 12-TRX, a dent-like depression in the profile around q = 0.4 nm^{-1} become less obvious and almost filled up at PEG = 9.5 M (39 wt %). The addition of PEG400 changed ρ_{sol} from 335 to 385 e/nm^3. For each profile, we made an attempt to fit all the data with same parameters with titrating [28]. The resultant theoretical curves are compared with the data in Figure 5b and obtained parameters are listed in Table 2. Except for the smaller undulation region, the agreement is good, indicating the 5th-layer model is good enough to describe the 12-TRX. The smaller undulation region reflects more detailed atomic structures and thus the simple multilayer model may be too course to reproduce these region. We may need a more elaborate method such as molecular dynamic calculation.

In order to confirm the parameters in Table 2, Figure 5c compares the calculated values with the data when other parameters were used. In Table 2, we conclude $t_1 = 20 \pm 1$nm. When we substitute $t_1 = 22$ nm at C_{PEG} = 0, it is clear that t_1 = 22 nm cannot reproduce the intensity around 0.03 nm$^{-1} < q <$ 0.4 nm^{-1}. In the case of $t_5 = 5$ at C_{PEG} =0, the theory did not agree with the data around 0.03 nm$^{-1} < q <$ 0.4 nm^{-1}. The inner headgroup density was smaller than that of the outer one by about 24 e/nm^3 for all samples. If we fitted in the condition of $\rho_2 = \rho_4$, the resultant curve did not fit the data in the same level of $\rho_2 \neq \rho_4$. The disagreement of them may be explained that the outer headgroup incorporates the phosphor-ester moiety of PEG-PE or inner headgroup incorporates the hydrophobic tails. The width of the headgroup and its electron density, and those of the alkyl domain take reasonable values in terms of their molecular structures [26]. Since PEG400 was added after liposome formation, we expected that PEG400 cannot enter the inner solvent compartment. For this case, we have to fix $\rho_1 = \rho_{PBS}$ for all fitting, however, if do so, the agreement became poor. We assume that the added PEG 400 could enter the inside, probably through fusion between liposomes. Although PEG 400 can be at inside, we cannot observed any evidence that PEG-PE anchors on the inner layer. Based on these above discussions, we can conclude this section that the addition of PEG 400 did not alter the liposome structure and only change the solvent electron density. By taking advantage, all of the structural parameters were uniquely determined.

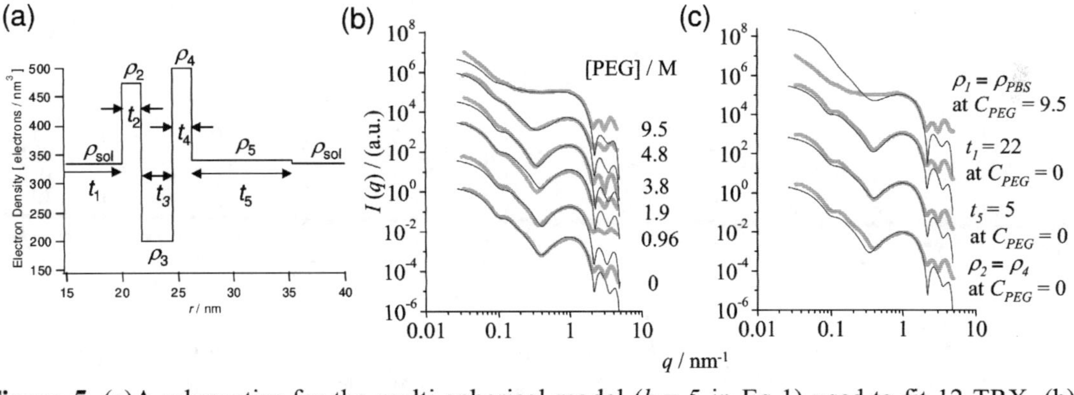

Figure 5. (a)A schematics for the multi spherical model (k = 5 in Eq 1) used to fit 12-TRX. (b) Changes of the scattering profiles of 12-TRX when PEG400 was added to solution at a fixed lipid concentration (gray lines) and their best fit curves (black line) calculated from Eq 1. (c) Theoretical curves when one of the fitting parameter was changed from the best fit combination, compared with the data.

Table 2. Characteristics of 12-TRX

Solvent	t_1	t_3	t_2 or t_4	t_5	ρ_2	ρ_3	ρ_4	ρ_5	ρ_{sol}
PBS	20	2.7	1.8	9	475	200	500	340	335
0.96M PEG, PBS	20	2.6	2	9	475	200	500	340	340
1.9M PEG, PBS	20	2.5	1.9	9	475	200	500	348	345
3.8M PEG, PBS	20	2.5	1.9	9	485	200	500	356	353
4.8M PEG, PBS	17	2.6	1.9	9	485	200	510	363	357
9.5M PEG, PBS	20	2.6	1.9	9	475	200	495	385	385

The units of t and ρ are nm and e / nm^3, respectively.

4-2 : Analysis of MLV of 0-TRX

To describe MLV (in our case two lamellae), we supposed $k = 7$ as presented in Figure 6a, where we assume the double lamella vesicle and the second and third headgroup domain are located close enough to give one layer for scattering. This assumption is reasonable since 0-TRX has no charged headgroup to induce repulsive interaction. We tried to fit the data with only the 5 th-layer model (Figure 6c, bottom), but the shoulder at $q = 1.0$ nm^{-1} was more enhanced than the experiments. According to the 7 th-layer model, the origin of the shoulder is diffraction between the three layers. These results indicates that 0-TRX consist of the mixtures of UMV and double MLV and the volume fraction is defined by ψ .

$$I(q) = \Psi\left[\sum_{i=1}^{5}\frac{V_i(\rho_i - \rho_{i+1})j_1(qR_i)}{qR_i}\right]^2 + (1-\Psi)\left[\sum_{i=1}^{7}\frac{V_i(\rho_i - \rho_{i+1})j_1(qR_i)}{qR_i}\right]^2 \quad (3)$$

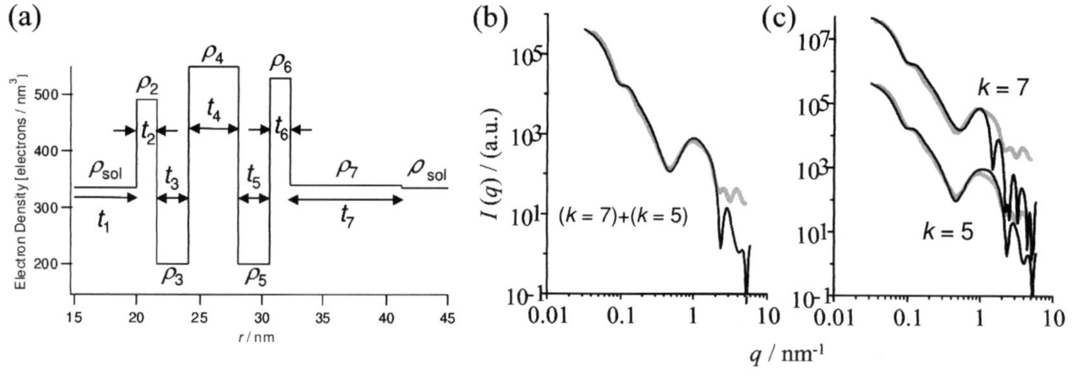

Figure 6. A model used to fit 0-TRX (a). Comparison of the data and theoretical curve calculated from Eq 3 that is a combination of $k = 5$ and $k = 7$ in Eq 3 (b). The panel (c) shows that the data can be fitted with neither $k = 5$ nor $k = 7$.

When we use this combinational equation, the experimental data was well fitted as presented in Figure 6b and the used parameters are listed in Table 3. Compared these values with Tabe 3, it can be seen that the outer and middle headgroup's electron densities for 0-TRX were larger than that of 12-TRX. This reason is not clear at this moment, but it probably due to our assumption that the second and third head domains were treated on layer. Even though this, the scattering profile was well reproduced with this model and we can conclude that this

Table 3. Characteristics of 0-TRX

Solvent	t_1	t_2 or t_6	t_3 or t_5	t_4	t_7	ρ_2	ρ_3 or ρ_5	ρ_4	ρ_6	ρ_7	ρ_{sol}
PBS	20	1.6	2.5	4	9	490	200	550	530	340	335

Solvent	t_1	t_2 or t_4	t_3	t_5	ρ_2	ρ_3	ρ_4	ρ_5	ρ_{sol}
PBS	24	2.5	1.6	9	490	200	530	340	335

The units of t and ρ are nm and e / nm^3, respectively.

combination model can describe the 0-TRX. According to our analysis, about $1-\psi = 0.071$ of liposomes took MLV.

4-3 : Other general discussions

DLS showed that 0-TRX always showed a larger particle size than 12- and 19-TRX, even all of them were filtered with the same pore size. When liposomes go through the small pore, their spherical shape might deform to prolate to go through the pore easily. MLV may be harder to deform than UMV, and thus after filtration UMV may be larger than MLV. Nevertheless, the result is opposite. This difference may be due to decrease of membrane flexibility by adding TRX to HSPC.

A grafted PEG chain on interface can classify two types [29]. One is isolated chain on the surface called "mushrooms" and the other is closely-spaced chains called "brush". The width of the PEG layer is 9 nm for our system. According to previous work [30], the radius of gyration of PEG 5000 is about 3.08 nm. By comparing these values, we can conclude the PEG on the liposome surface is in the brush state. From the ρ_5 and ρ_{PEG}, we can evaluate the PEG volume percent in the PEG layer and roughly estimated how many PEG-PE molecules anchored in one liposome. This average number is estimated to be about 850.

5. Conclusions

The present paper examined the structural changes of the PEG-liposomes when TRX was added. TEM showed the presence of ULV liposomes with a diameter around 100nm for all the compositions. These results were consisted with the DLS results. TEM also indicated that 0-TRX liposomes contained some MLVs, while 12- and 19-TRX did not. When we analyzed the SAXS profiles from 12-TRX combined with the contrast variation technique, the SAXS data was consistently explained with a ULV model. For 0-TRX, when we assumed a mixture of ULV and MLV, the data could be reconstructed.

Journal of Physics: Conference Series **272** (2011) 012011 doi:10.1088/1742-6596/272/1/012011

References

[1] Pandit SA, Bostick D, Berkowitz ML 2003 *Biophys. J.* **85** 3120
[2] Gradzielski M 2003 *J. Phys.: Condens. Matter* **15** R655
[3] Tresset G 2009 *PMC biophysics* **2** 3
[4] Kloepfer JA, Cohen N, Nadeau JL 2004 *J. Phys. Chem. B* **108** 17042
[5] Wijaya A, Hamad-Schifferli K 2007 *Langmuir* **23** 9546
[6] Allen TM 1994 *Trends Pharmacol Sci.* **15** 215
[7] Woodle MC 1995 *Adv. Drug. Del. Rev.* **16** 249
[8] Mobed M, Chang TM 1998 *Biomaterials* **19** 1167
[9] Sakai H, Takeoka S, Park SI, Kose T, Nishide H, Izumi Y, et al. 1997 *Bioconjug Chem.* **8** 23
[10] Bradley AJ, Devine DV, Ansell SM, Janzen J, Brooks DE 1998 *Arch. Biochem. Biophys.* **357** 185
[11] Kenworthy AK, Hristova K, Needham D, McIntosh TJ 1995 *Biophys. J.* **68** 1921
[12] Zalipsky S 1995 *Bioconjug Chem.* **6** 150
[13] Koiwai K, Tokuhisa K, Karinaga R, Kudo Y, Kusuki S, Takeda Y, et al. 2005 *Bioconjug Chem.* **16** 1349
[14] Sakuragi M, Kusuki S, Hamada E, Masunaga H, Ogawa H, Akiba I, et al. 2009 *J. Phys.: Conference series* **184** 012008.
[15] Sakurai K, Kusuki S, Hamada E, Takeda Y. JPA 2008-143844.
[16] Harigai T, Kondo M, Isozaki M, Kasukawa H, Hagiwara H, Uchiyama H, et al. 2001 *Pharm. Res.* **18** 1284
[17] Kawahara K, Sekiguchi A, Kiyoki E, Ueda T, Shimamura K, Kurosaki Y, et al. 2003 *Chem. Pharm. Bull.* **51** 336
[18] Garbuzenko O, Zalipsky S, Qazen M, Barenholz Y. 2005 *Langmuir* **21** 2560
[19] Shimizu K, Isozaki M, Koiwai K. WO97/42166
[20] Sakurai K 2008 Small-Angle X-ray Scattering from Ultra-Dilute Solutions of pDNA/Cationic Lipid Complexes. X-ray and Neutron Techniques for Nano-Structural Research. SPring-8, Hyogo, Japan
[21] Guinier A and Fournet G 1955 *Small-Angle Scattering of X-rays* (Inc.: New York)
[22] Hirai M, Iwase H, Hayakawa T, Koizumi M, Takahashi H 2003 *Biophys. J.* **85** 1600
[23] Kawaguchi T, Hamanaka T, Kito Y, Machida H 1991 *J. Phys. Chem.* **95** 3837
[24] Bolze J, Horner KD, Ballauff M 1996 *Langmuir* **12** 2906
[25] Roe RJ 2000 Methods of X-ray and neutron scattering in polymer science. pp 17-22 (Oxford University Press, New York.)
[26] Nagle JF, Tristram-Nagle S. 2000 *Biochim.Biophys. Acta.* **1469** 159
[27] Eguchi K, Kaneda I, Hiwatari Y, Masunaga H, Sakurai K 2007 *J. Appl. Crystallogr.* **40** s264
[28] Naruse K, Eguchi K, Akiba I, Sakurai K, Masunaga H, Ogawa H, et al. 2009 *J. Phys. Chem. B* **113** 10222
[29] Tirosh O, Barenholz Y, Katzhendler J, Priev A 1998 *Biophys. J.* **74** 1371
[30] Devanand K and Selser JC 1991 *Macromolecules* 1991 **24** 5943

Journal of Physics: Conference Series **272** (2011) 012012 doi:10.1088/1742-6596/272/1/012012

Preparation and Characterization of a pH-Responsive Core Cross-linked Polymer Micelle

Shouta Kousaka, Makoto Sugahara, Tatsuya Endo, and Shin-ichi Yusa[1]
Department of Materials Science and Chemistry, Graduate School of Engineering, University of Hyogo, 2167 Shosha, Himeji, Hyogo 671-2280, Japan.

E-mail: yusa@eng.u-hyogo.ac.jp

Abstract. Poly(ethylene glycol)-*b*-poly(2-(diethylamino) ethyl methacrylate-*co*-2-cinnamoyloxyethyl acrylate) (PEG-*b*-P(DEA/CEA)) was prepared by reversible addition–fragmentation chain transfer (RAFT) controlled radical polymerization. pH-responsive association behaviour of PEG-*b*-P(DEA/CEA) in 0.1 M NaCl was characterized by dynamic light scattering (DLS). As solution pH is increased from an acidic pH, the hydrodynamic radius (R_h) increases, indicative of the polymer micelle formation. The formation of a micelle was also supported by static light scattering (SLS) data. The cinnamoyl groups in the core of the polymer micelle undergo photodimerization, yielding cross-links between polymer chains. The core of the polymer micelle was fixed, which was confirmed by DLS, SLS, and small angle X-ray scattering (SAXS) techniques. When pH is decreased to 3, R_h of the core cross-linked (CCL) polymer micelle slightly increases due to the protonation of the DEA unit in the cross-linked core. The reversible pH-induced swelling and shrinking behaviour can be observed.

1. Introduction

Stimuli-responsive polymer micelles composed of amphiphilic diblock copolymers have received considerable attention as a candidate for controlled delivery applications, including drug delivery system (DDS) [1-3]. However, if polymer micelles are used for this purpose, they may experience dilution below the critical micelle concentration (cmc), leading to dissociation into unimers. In such a case, the encapsulated guest molecules cannot be delivered to the target. On the other hand, core cross-linked (CCL) polymer micelles are expected to have a much higher stability of the encapsulation of guest molecules at a diluted concentration than uncross-linked polymer micelles.

We prepared a stimuli-responsive CCL micelle in water by a photo-cross-linking technique using diblock copolymer micelles. We reported on the preparation of a pH-responsive copolymer (Scheme 1), poly(ethylene glycol)-*b*-poly(2-(diethylamino) ethyl methacrylate-*co*-2-cinnamoyloxyethyl acrylate) (PEG-*b*-P(DEA/CEA)). PEG-*b*-P(DEA/CEA) was synthesized from precursor polymer (PEG-*b*-P(DEA/HEA)) prepared via reversible addition-fragmentation chain transfer (RAFT) radical polymerization using PEG based chain transfer agent (PEG-CTA) [4-6]. The PEG block is soluble in water independent of pH. On the other hand, the solubility of DEA depends on pH. CEA is the photo-cross-linkable unit. A CCL micelle was prepared by UV-irradiation. The CCL micelle will be expected for the application of its solubilizing ability for hydrophobic molecules and controlled release by changing pH.

[1] To whom any correspondence should be addressed.

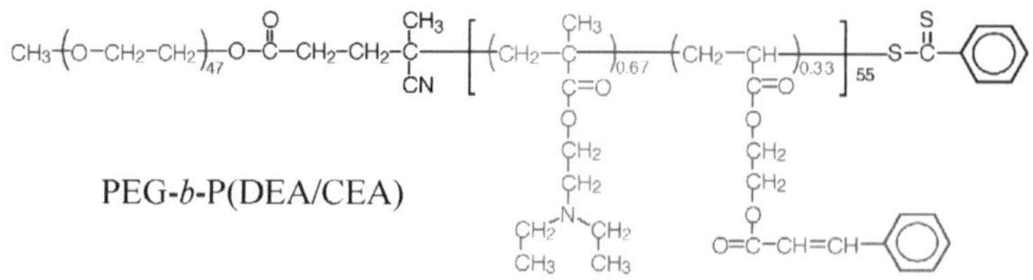

Scheme 1. Chemical structure of PEG-*b*-P(DEA/CEA)

2. Experimental

2.1. *Materials.* DEA was passed through basic alumina columns to remove inhibitor. 2-Hydroxyethyl-acrylate (HEA) was dried and purified by distillation under reduced pressure. PEG-CTA was synthesized according to a modified literature procedure [7].

2.2. *Preparation of Precursor Diblock Copolymer.* Precursor diblock copolymer (PEG-*b*-P(DEA/HEA)) was prepared by RAFT polymerization. DEA (3.08 g, 16.6 mmol) was dissolved in 30 mL of water, and the solution pH was adjusted to 7 using HCl (6.0 M). PEG-CTA (750 mg, 0.316 mmol), HEA (1.54 g, 13.3 mmol), and 4,4'-azobis (4-cyanopentanoic acid) (27.9 mg, 0.0995 mmol) were added to the solution. Polymerization was carried out at 70 °C for 3 h under Ar atmosphere. The block copolymer was purified by dialysis against pure water. PEG-*b*-P(DEA/HEA) was recovered by a freeze-drying technique (yield 3.71 g). Number-average molecular weight (M_n) and molecular weight distribution (M_w/M_n) determined by gel-permeation chromatography (GPC) are 1.00×10^4 and 1.09, respectively. The total number-average degree of polymerization (DP) for DEA and HEA is 55, and the molar contents of DEA and HEA units in the P(DEA/HEA) block are 67 and 33 mol %, respectively, estimated by ^{1}H NMR. To investigate relationship between M_n and conversion, predetermined amounts of neutralized DEA, HEA, PEG-CTA, and initiator were dissolved in D_2O. The solution was heated at 70 °C under Ar atmosphere for varying lengths of time. Conversion and M_n were characterized by ^{1}H NMR and GPC, respectively.

2.3. *Introduction of Cinnamoyl Group to PEG-b-P(DEA/HEA).* Cinnamoyl chloride (1.10 g, 6.60 mmol) and PEG-*b*-P(DEA/HEA) (3.20 g, 0.283 mmol) were dissolved in pyridine (28.0 g, 0.354 mol), and the mixture was stirred for 22 h at room temperature. The mixture was poured into a large excess of diethyl ether to precipitate the polymer. PEG-*b*-P(DEA/CEA) was recovered (yield 3.34 g). The conversion of HEA to CEA was almost completed confirmed by ^{1}H NMR.

2.4. *Micelle Cross-Linking.* PEG-*b*-P(DEA/CEA) was dissolved in a 0.1 M NaCl aqueous solution, and the pH was adjusted to 3 or 10, where the polymer concentration (C_p) is 5.0 g/L. The photo-dimerization of cinnamoyl groups was carried out by using an Asahi Spectra MAX-300 equipped with 300-W Xe lamp and a 275-nm cut-off filter. The light intensity was 20 mW/cm^2 at 350 nm. The change in absorption spectra of PEG-*b*-P(DEA/CEA) during photoirradiation was monitored by UV-visible absorption spectra.

2.5. *Measurements.* GPC analysis was performed using with a SB-804 HQ column working at 40 °C under a flow rate of 0.6 mL/min. A 0.3 M Na_2SO_4 aqueous solution containing a 0.5 M acetic acid was used as eluent. ^{1}H NMR spectra were obtained with a Bruker DRX-500 spectrometer. Light scattering data were obtained at 25 °C with an Otsuka Electronics Photal DLS-7000HL equipped with an ALV5000/EPP multi-τ digital time correlator. A He-Ne laser (10.0 mW at 632.8 nm) was used as a light source. Sample solutions for light scattering measurements were prepared with 0.1 M NaCl, which were filtered with a 0.2 μm pore size membrane filter. In dynamic light scattering (DLS) measurements, to obtain the relaxation time distribution, $\tau A(\tau)$, the inverse Laplace transform (ILT) analysis was performed using the algorithm REPES [8]. The hydrodynamic radius (R_h) is calculated using the Einstein–Stokes relation. In static light scattering (SLS) measurements, the angular dependence of the excess absolute time-averaged scattered intensity, also known as the Rayleigh ratio $R_{vv}(\theta)$, was measured [9]. SAXS measurements were carried out at the BL40B2 of SPring-8, Harima, Japan using an incident X-ray with the wavelength, $\lambda = 0.150$ nm. The measured sample was 5.0 g/L of the CCL micelle in a 2 mm glass capillary. The SAXS scattering intensity from isolated particle in solution was used to model fitting to ignore the contribution of interparticle interaction. The hard sphere and spherical core-shell models were used to analyze the scattering intensity from the CCL micelle [10, 11].

3. Results and Discussion

We studied dependence of M_n and M_w/M_n on the average conversion for the random copolymerization of DEA and HEA using PEG-CTA via RAFT (Figure 1). M_n increased almost linearly with the conversion, while M_w/M_n remained nearly constant independent of conversion. All the values of M_n are close to the theoretical values predicted from the living mechanism.

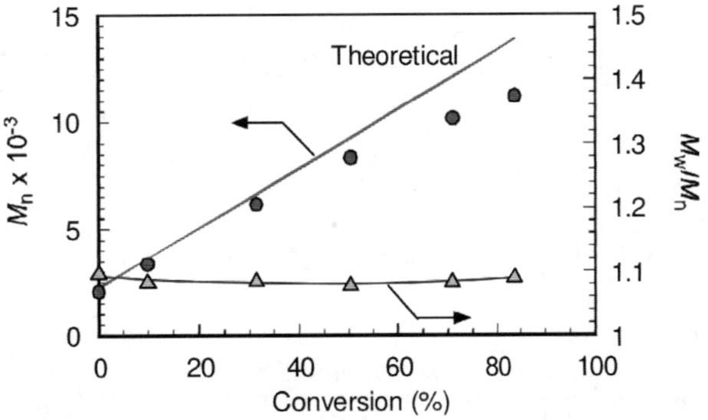

Figure 1. Dependence of M_n and M_w/M_n on the monomer conversion in the polymerization of DEA and HEA in the presence of PEG-CTA in D_2O at 70 °C under Ar atmosphere.

Table 1. Light Scattering Data for Unimer, Polymer Micelle, and CCL Micelle

Sample	pH	R_h (nm)	M_w x 10^4	N_{agg}
Unimer	3	4.0	3.5	-
Micelle	10	12.5	320	92
CCL micelle	3	13.6	340	98

Before UV-irradiation, the R_h and M_w values of PEG-*b*-P(DEA/CEA) at pH 3 were 4.0 nm and 3.5 × 10^4, respectively, indicated that the polymer was dissolved as a unimer. On the other hand, at pH 10 the R_h and M_w values of the polymer were 12.5 nm and 3.2×10^6 , respectively, indicated that the polymer aggregated to form micelle. At pH 10, the photodimerization of the pendent cinnamoyl groups leads to the cross-linking of different polymer chains in the micelle core. The R_h value for the CCL micelle at pH 10 was found to be the same as that of the un-cross-linked polymer micelle at pH 10. When the solution pH was decreased to 3, the R_h value for the CCL micelle increased from 12.5 to 13.6 nm, which suggests that the CCL micelle remains. The CCL micelle core is swollen in acidic solutions due to the protonation of the DEA units in the cross-linked core. The pH-induced reversible swelling and shrinking of the CCL micelle was observed. The values of aggregation number (N_{agg}) can be calculated from the ratio of weight-average molecular weight (M_w) for the polymer micelle or the CCL micelle and the unimer. The

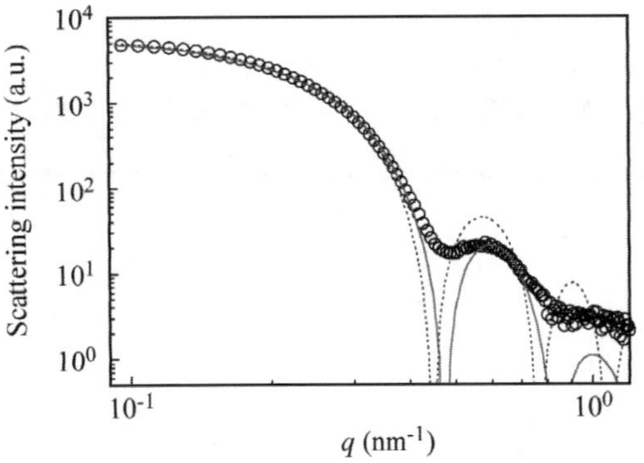

Figure 2. SAXS data for a CCL micelle aqueous solution containing 0.1 M NaCl at pH 3. The data were fitted to the monodisperse hard sphere model (---) or to the monodisperse spherical core-shell model (—).

light scattering data for PEG-*b*-P(DEA/CEA) as a unimer at pH 3, the polymer micelle at pH 10, and the CCL micelle at pH 3 are summarized in Table 1.

Figure 2 shows SAXS data for the CCL micelle in 0.1 M NaCl at pH 3. The data were examined by fitting to a hard sphere model and a spherical core-shell model. The data are reasonably well fitted to the spherical core-shell model over the entire q range, whereas the hard sphere model does not yield a good fit. By fitting the SAXS data to the spherical core-shell model, radius of core and thickness of shell values were estimated to be 9 and 3 nm, respectively. This result indicates that the core of cross-linked P(DEA/CEA) is surrounded by PEG shell. The overall radius of the CCL micelle estimated from SAXS data (= 12 nm) is fairly close to that estimated from light scattering data.

4. Conclusion

PEG-*b*-P(DEA/HEA) was prepared by RAFT polymerization using PEG-CTA followed by chemical modification into PEG-*b*-P(DEA/CEA). PEG-*b*-P(DEA/CEA) exhibited reversible micellization and dissociation induced by changing solution pH. The polymer formed micelles at pH 10. The core of the polymer micelle can be cross-linked by UV-irradiation at pH 10. The R_h values for the CCL micelle at pH 3 and 10 were 12.5 and 13.6 nm, respectively. The CCL micelle indicated reversible swelling and shrinking by changing solution pH. The SAXS data suggested that the CCL micelle composed of cross-linked pH-responsive core and PEG shell.

Acknowledgments

Dr. Sono Sasaki and Dr. Hiroyasu Masunaga of the Japan Synchrotron Radiation Research Institute are acknowledged for their help on SAXS measurement at SPring-8 (Proposal No. 2007A1591). This work was financially supported by a Grant-in-Aid for Scientific Research from the Japanese government (No.22350106).

References

[1] Zhang H, Mardyani S, Chan W C W and Kumacheva E 2006 *Biomacromolecules* **7** 1568
[2] Oh J K, Siegwart D J and Matyjaszewski K 2007 *Biomacromolecules* **8** 3326
[3] Missirlis D, Tirelli N and Hubbell J A 2005 *Langmuir* **21** 2605
[4] Li Y, Lokitz B S and McCormick C L 2006 *Macromolecules* **39** 81
[5] Shi L, Chapman T M and Beckman E J 2003 *Macromolecules* **36** 2563
[6] Chong Y K, Le T P T, Moad G, Rizzardo E and Thang S H 1999 *Macromolecules* **32** 2071
[7] Yusa S, Yokoyama Y and Morishima Y 2009 *Macromolecules* **42** 376
[8] Schillén K, Brown W and Johnsen R M 1994 *Macromolecules* **27** 4825
[9] Zimm B H 1948 *J. Chem. Phys.* **16** 1099
[10] Pedersen J S 1997 *Adv. Colloid Interface Sci.* **70** 171
[11] Guiner A and Fournet G Eds 1995 *Small Angle Scattering of X-rays* (John Wiley & Sons)

Surface Segregation of Nanoparticles Driven by Supercritical Carbon Dioxide

Mitsunori Asada[1,2], Peter Gin[2], Maya K. Endoh[2], Sushil K. Satija[3] and Tadanori Koga[2,4]

[1]Kurashiki Research Center, Kuraray Co., Ltd., Okayama 710-0801, Japan, [2]Depertment of Materials Science and Engineering, Stony Brook University, Stony Brook, New York 11794-2275, [3]Center for Neutron Research, NIST, Gaithersburg, Maryland 20899, [4]Chemical and Molecular Engineering Program, Stony Brook University, Stony Brook, New York 11794-2275

E-mail: tkoga@notes.cc.sunysb.edu

Abstract. Surface segregation (i.e., preferential segregation of one component to the surface in multicomponent systems) is common to all material classes and is typically driven by a reduction in surface energy which compensates for the entropy loss and/or energy gain associated with the demixing of the components. However, the conventional surface segregation for polymeric systems requires high temperatures, typically close to 200°C, and long annealing time, in order to ensure enough polymer mobility. Here we show a low-temperature method to preferentially migrate organoclay nanoparticles to the polymer surface using supercritical carbon dioxide.

1. Introduction

The self-assembly of nanoparticles is an attractive approach for nanostructure fabrication. The control of layered structures on the nanometer scale is particularly desirable for thin-film organic devices, because this essentially affects the electrical, optical, and mechanical properties. To spontaneously fabricate such layered structures, careful control of driving forces such as phase separation, crystallization, or surface energy of materials is necessary. Krishan et al. reported a new means to obtain self-assembled multilayer structures at the nanometer level: the counterbalance between the conformational entropy gain of moving the linear polymer chain away from the substrate and mixing enthalpy and translational entropy losses of the nanoparticles by moving to the substrate interface [1]. Recently, Wei et al. synthesized a fullerene (C_{60}) derivative with a fluorocarbon chain, which has a low surface energy and prefers to migrate to the air/polymer interface during spin-coating process, and could create a very thin layer of the nanoparticles at the polymer/air interface to be used for organic thin film transistors [2]. Although these studies propose new attractive methods for fabrication of various organic thin-film devices, their applications to actual electronic devices are still very limited and remain challenging. In this paper, with the aim of developing a new route to control the self-assembling process of nanoparticles embedded in polymer thin films, we describe simple and effective surface modifications applicable to a wide variety of polymer nanocomposite films using supercritical carbon dioxide ($scCO_2$, the critical point is T_c = 31.3 °C and P_c = 7.38 MPa). $ScCO_2$ is being used increasingly as an environmentally friendly solvent for polymer processing and synthesis [3]. The novel temperature and pressure conditions near the critical point where the anomalous absorption of CO_2 molecules into "CO_2-phobic" polymer thin films [4] were utilized in the study.

2. Experimental

A polymer examined for this study was monodisperse polystyrene (PS, Pressure Chemical Co., weight average molecular weight (M_w) = 200,000, polydispersity index = 1.06). Organoclay particle (Cloisite 6A, a natural montmorillonite with alkyl *N*, *N*-dimethyl dehydrogenated tallow quaternary ammonium chloride) was used as a model. The nanoparticle was firstly dissolved in toluene and the polymer was then added to the solution. The weight fraction of the clay nanopatcicles against the polymer was 0.01. The solutions were sonicated for 60 seconds prior to spin-coating process in order to prevent from aggregation of the nanoparticles. The surface free energies of PS and clay are 33 and 30-35 mJm^{-2}, respectively. The polymer nanocomposite films without pre-annealing at high temperature were then placed in the high pressure cell [5] and immersed in $scCO_2$ under the unique condition of T = 36 °C and P = 8.2 MPa, where the anomalous swelling of various polymer thin films was induced by long-range density fluctuations (hereafter we call the T and P condition as the "ridge" condition). We also examined the surface structures of the pre-annealed films at T = 170 °C in vacuum for 2 h as controls. Surface morphologies of all the films were observed by an atomic force microscope (AFM) and lateral force microscope (LFM) (Digital Instruments, Dimension 3100) with a silicon nitride cantilever. The images obtained were then analyzed by using ImageJ 1.40g software (NIH, USA).

3. Results and Discussion

Fig. 1(a) and 1(b) present the topographic surface morphologies of the as-spun and thermally annealed PS/clay thin films measured by AFM. We can see that the film annealed at T = 170 °C in vacuum for 2 h has a rough surface, while the surface of the original as-spun film is relatively flat. Fig. 1(d) and 1(e) correspond to the cross-sectional profiles of Fig. 1(a) and 1(b), respectively, indicating that the dewetting takes place during the annealing process. From the X-ray reflectivity measurement for the as-spun PS/clay film, we observed the monolayer formation of the clay particles at the substrate, indicating that the strong affinity of the clays for a native oxide layer and the dewetting is caused by the poor interaction between the clay particle and PS [6]. Fig. 1(c) and 1(f) are the surface morphologies of the $scCO_2$ treated PS/clay thin film at the ridge condition and its cross-sectional profile, respectively. Interestingly, no dewetting was observed even after 23 h exposure time in $scCO_2$ and rather the surface became rougher due to the segregation of the clay particle (the bright spots in the AFM images correspond to the clay nanoparticles).

In order to quantitatively characterize the surface segregation structures of the clay nanoparticles, we conducted an image analysis on the LFM pictures, which offer a frictional force contrast on a surface. This

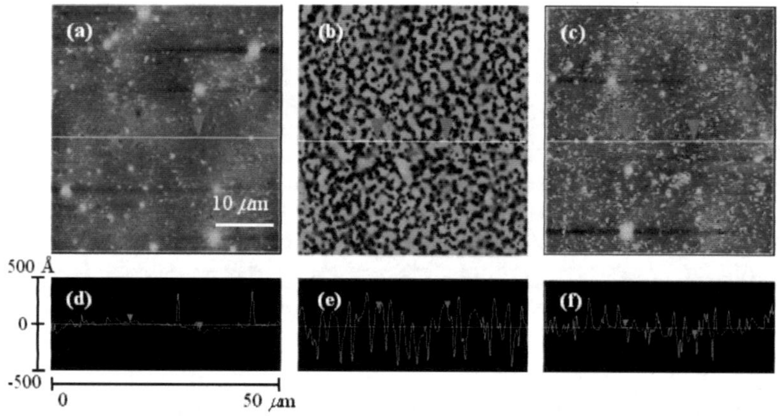

Figure 1. Topographic images for the (a) as-spun, (b) thermally annealed (T = 170 °C for 24 h in vacuum) and (c) CO_2 treated (P = 8.2MPa and T = 36 °C for 1 h) PS/clay thin films. The scan sizes and height scales for the images are 50×50 μm^2 and 0-1000 Å, respectively. (d), (e) and (f) show the cross-sectional profiles along the white lines in (a), (b) and (c), respectively. The height scales for these profiles are 0-1000 Å.

allows us to quantify the surface exposing clay nanoparticles with relatively large difference in the frictional nature. The calculated areas of the surface exposing clays in the PS/clay thin film (both 50 and 100 nm thickness) are presented in Fig. 2. For the as-spun films of both 50 and 100 nm in thick, approximately 7.5 % in topmost area was covered by the clay particles. On the other hand, the surface coverage of the particles after exposing to $scCO_2$ increased up to more than 10 % for both films, which corresponds to an approximately 40 % increase from the unexposed film. Hence, it is obvious that the $scCO_2$ treatment induces the surface segregation of the particles. From the figure we see no significant time dependence of the surface exposed clay nanoparticles after 1 h exposure, suggesting the surface segregated structure is an equilibrium state. In addition, we also found that the observed surface segregation is very sensitive to employed *T* and *P* conditions; it can take place only at the ridge condition, indicating that the density fluctuating $scCO_2$ is a key for the phenomenon.

We also investigated polyhedral oligomeric silsesquioxane nanoparticles co-polymerized with styrene monomers (the weight fraction of the POSS component was 15%) embedded in the same PS matrix by using AFM and neutron reflectivity and found similar surface segregation of the nanoparticles after the same $scCO_2$ treatment. Hence, it is clear that the enthalpic interactions are not the central driving force for the $scCO_2$-induced surface migration. Rather, an entropic effect imposed on the polymer chains owing to the presence of the nanoparticles would be crucial. As the individual polymer chains must extend and stretch around the nanoparticles to achieve the equilibrium swelling in $scCO_2$, there is an entropic gain in expelling the nanoparticles to the free surface, as previously reported for surface segregation of nanoparticles at the air/polymer interface [7]. Another important factor would be the critical wetting of CO_2 molecules onto nanoparticle surfaces in the highly compressible region [8]. These combined effects could push the nanoparticles out of the film even though the nanoparticles are 'soluble' in the polymer matrix.

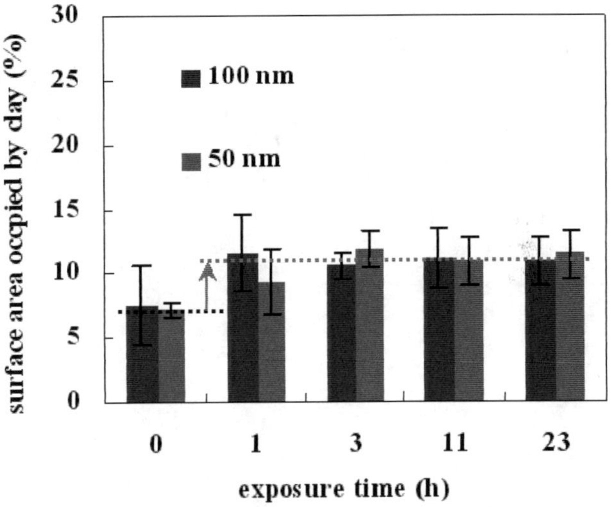

Figure 2. Calculated surface areas occupied by the clay nanoparticles as a function of CO_2 exposure time.

4. Conclusion

In summary, we have described a very simple, yet robust approach to induce the surface segregation of nanoparticles by using supercritical CO_2 under the unique *T* and *P* condition. The resultant surface structures can be then preserved by the vitrification of the polymer chains via rapid pressure quenching. Hence, this experimental finding would facilitate the use of $scCO_2$ as a low-temperature and environmentally friendly solvent for creating a variety of nano-patterned polymer surfaces tailored to one's needs.

Acknowledgements
This work was supported by NSF Grant No. CMMI-0846267 and Kuraray Co., Ltd., Japan.

References

[1] Krishnan R S, Mackay M E, Hawker C J and Horn B V 2005 *Langmuir* **21** 5770.
[2] Wei Q, Nishizawa T, Tajima K and Hashimoto K 2008 *Adv. Mater.* **20** 2211.

[3] McHugh M A and Krukonis V J 1994 *Supercritical Fluid Extraction: Principles and Practice* (Butterworth-Heinemann, Boston).
[4] Koga T, Seo Y S, Zhang Y, Shin K, Kusano K, Nishikawa K, Rafailovich M H, Sokolov J C, Chu B, Peiffer D, Occhiogrosso R and Satija S K 2002 *Phys. Rev. Lett.* **89** 125506.
[5] Koga T, Seo Y S, Shin K, Zhang Y, Rafailovich M H, Sokolov J C, Chu B and Satija S K 2003 *Macromolecules* **36** 5236.
[6] Si M, Araki T, Ade H, Kilcoyne A L D, Fisher R, Sokolov J C and Rafailovich M H 2006 *Macromolecules* **39** 4793.
[7] Tyagi S, Lee J Y, Buxton G A and Balazs A C 2004 *Macromolecules* **37** 9160.
[8] Findenegg G H 1983 *In Fundamentals of Adsorption* (Engineering Foundation, New York) 207-218.

Journal of Physics: Conference Series **272** (2011) 012014 doi:10.1088/1742-6596/272/1/012014

Improvement of SAXS Measurement near the Sulfur *K*-edge

Masashi Handa[1], Yuya Shinohara[1,4], Hiroyuki Kishimoto[1,2], Yusuke Tamenori[3], Naoto Yagi[3,4], and Yoshiyuki Amemiya[1,4]
[1] Department of Advanced Materials Science, Graduate School of Frontier Sciences, The University of Tokyo, 5-1-5 Kashiwanoha, Kashiwa, Chiba, Japan, [2] Sumitomo Rubber Industries Co. Ltd., Kobe, Hyogo 651-0071, Japan, [3] JASRI/SPring-8, 1-1-1 Kouto, Sayo 679-5198, Japan, [4] CREST, JST Japan.

E-mail: yuya@k.u-tokyo.ac.jp

Abstract. We have made feasibility study of anomalous small-angle X-ray scattering (ASAXS) near the sulfur *K*-edge (2.472 keV). In the present study, we report the improvement of SAXS measurement with the use of Si (111) channel-cut monochromator and Kratky camera at BL27SU, SPring-8. The photon flux and the energy resolution of the incident X-ray beam increased by 10 times and 2.5 times, respectively. SAXS intensity also increased by about 10 times with the improved low-q and high-q limits, enabling us to measure SAXS with a better S/N ratio over a wider q range.

1. Introduction

Small-angle X-ray scattering (SAXS) is greatly effective in analyzing the structure of soft matter (for example [1]). The contrast of scattering length density, which causes SAXS, can be varied with anomalous dispersion. The scattering length of a constituent element, *f*, has a dependence on the energy of incident X-rays near its absorption edge as

$$f(E) = f_0 + f'(E) + if''(E)$$

where f' and f'' represent the real and imaginary parts of the dispersion corrections, respectively. By using the anomalous dispersion one can get the structural information related to the specific element. SAXS using this dispersion is called as anomalous SAXS (ASAXS). ASAXS has been widely applied to the systems containing high-*Z* elements whose *K*-edges are located in a hard X-ray region ($E > 4$ keV), while it has been poorly explored in a soft X-ray region ($E < 4$ keV). There are, however, some buried treasures of *K*-edges of important elements in biological and synthetic macromolecules such as silicon, phosphorus, and sulfur [2]. The *K*-edge of sulfur, the main target of the present study, is located at $E = 2.472$ keV.

Recently, we have made feasibility study of SAXS and ASAXS near the sulfur *K*-edge [3]. In the previous study, we used a varied-line-spacing plane grating monochromator (VLS-PGM), which had been originally equipped at BL27SU, SPring-8 [4, 5]. The flux available with this monochromator was in the order of 10^{11} photons/s/100mA/0.03%B.W. in an energy range of 0.10 - 2.0 keV, but it decreased by two orders of magnitude in an energy range above 2.0 keV [6]. This is mainly because the four mirrors and grating in the previous setting were all coated with gold having its *M*-edges at 2.21 and 2.29 keV, and X-ray beam was greatly absorbed by these optical elements.

In the present study, we have arranged a new experimental setting at BL27SU that is equipped with Si channel-cut monochromator and mirrors coated with Ni so as to increase the photon flux and the energy resolution. The characteristics of the X-ray beam and SAXS intensity profiles with the new setting were examined and compared to those in the previous study.

2. Experimental

We performed the experiment at B-branch of BL27SU, SPring-8 (Hyogo, Japan). All optical elements were arranged as shown in Fig. 1. In this layout, an X-ray beam was diverted into the experimental station by pre-mirror (M0). Three slits (S1, S2, and S3) were used to collimate the X-ray beam. Kirkpatrick-Baez (KB) mirrors were temporarily adopted to focus the X-ray beam at the sample position where the beam size was about 10 μm × 10 μm. These mirrors were coated with Ni. Between horizontal and vertical mirrors, Si (111) channel-cut monochromator was placed to monochromatize the X-ray beam. By the use of this monochromator, the X-ray beam position shifted in the vertical direction when the energy was changed. We set a photodiode (AXUVPSV, International Radiation Detectors Inc.) adjustable in the vertical direction just downstream of the sample to measure the photon flux. This photodiode also played an important role as a beamstop. The combination of S2, S3 and the photodiode was arranged in the manner of Kratky camera [7], which efficiently blocked parasitic scatterings in a low-q range. All these elements were placed in a vacuum chamber. It had a small aperture of 2 mm in diameter between the vertical mirror and the third slit, and the X-ray beam passed through the aperture. This aperture limited the range of available energy to about 200 eV. The installation of a fixed-exit X-ray monochromator is now planned to overcome this problem.

An indirectly illuminated X-ray area detector was used for SAXS measurement. This area detector was composed of a phosphor screen (P43, Gd_2O_2S:Tb), an optical relay lens, and a full-frame back-illuminated CCD detector (C4742-98-KAG, Hamamatsu Photonics Co. Ltd.). The details of the detector were described in the previous report [4].

For SAXS measurement, we used a nanocomposite as a model sample. The matrix was thiokol (LP-2, Toray Fine Chemicals Co. Ltd.) whose chemical formula was $-C_2H_4$-O-CH_2-O-C_2H_4-S-S-. Monodispersive spherical silica particles (Seahoster, Nippon Shokubai Co. Ltd.) of 130 nm in diameter were added to thiokol. The volume fraction of the particles was 6.0 %. After the particles were added to thiokol, the sample was stirred and kept at 120°C for 72 hours so that it stiffened. Then it was sliced into a piece of 30 μm in thickness by using a microtome.

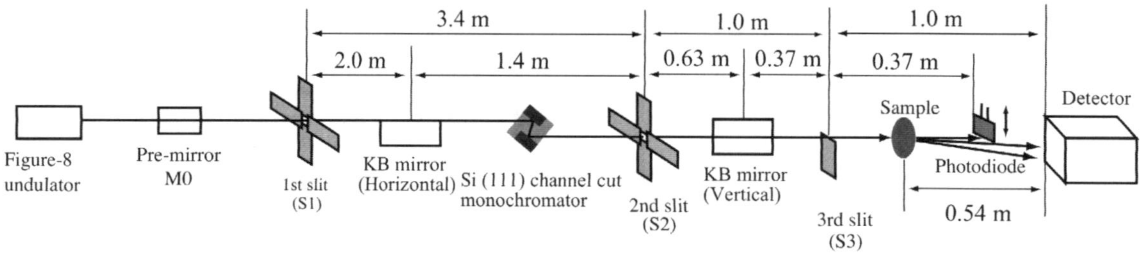

Figure 1. Schematic view of the experimental setting at B-branch of BL27SU. The third slit, sample and photodiode are placed in a differential pumping system [8].

3. Results and Discussion

3.1 Measurement of photon flux

We measured the photon flux by using the photodiode under the condition that the sample and all slits were removed; the result was shown in Fig. 2. We can see that the flux measured with the present setting (solid line) is more than 10^{10} photons/s/100mA/0.03%B.W. near the sulfur *K*-edge (2.472 keV), which is favorably compared to that with the previous setting (dotted line). The available energy range was limited to about 200 eV by the aperture of the vacuum chamber. It should be noted that in this experiment a front-end slit was not fully opened. Accordingly there is still room for obtaining a higher flux by fully opening the front-end slit. In this setting, the energy resolution $\Delta E/E$ has also improved, which is calculated as

$$\Delta E / E = \sqrt{\omega^2 + \varphi^2} \cot \theta_B$$

where ω is a full width at half maximum (FWHM) of the rocking curve of Si 111 reflection, φ is the angular divergence of the incident X-ray beam, and θ_B is the Bragg angle. According to the data of BL27SU, ϕ is 0.0108 mrad. Assuming that E = 2.5 keV and Debye-Waller factor is vanishingly small, ω is calculated to be 0.1724 mrad [9]. Thus, $\Delta E/E$ with this setting is calculated to be 1.3×10^{-4}. This is 2.5 times higher than that with the previous setting.

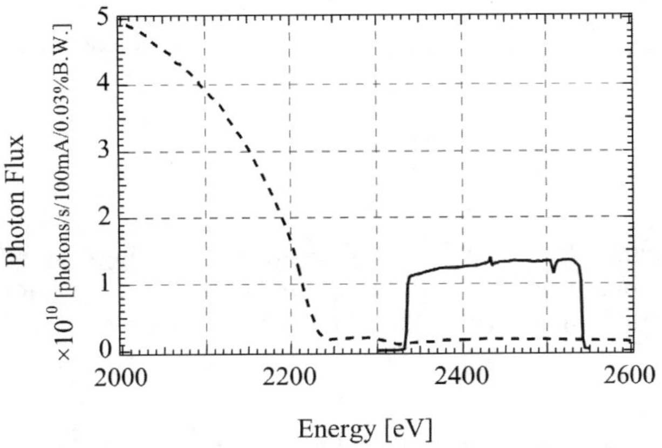

Figure 2. Photon flux with the two different settings. Solid line shows the flux with the improved setting (Si channel-cut monochromator) and dotted line shows that with the previous setting (the combination of grating, slits, and mirrors, VLS-PGM). They are normalized to a ring current of 100 mA and $E/\Delta E$ = 3000.

3.2 SAXS measurement.

We performed SAXS measurement in a transmission mode under the condition that all slits were set to collimate the X-ray beam. SAXS pattern was taken at E = 2460 eV. From the two-dimentional SAXS pattern, we calculated one-dimentional intensity profile by circularly averaging the intensities of corresponding pixels, and the result is shown together with the previous result in Fig. 3. The profiles need to be normalized with some factors because of the different experimental and sample conditions; the exposure time in the present study was 200 s, whereas 300 s in the previous study. In addition, the sample used in the present study differed in the thickness and the volume fraction of nanoparticles. Then the obtained profiles were normalized in terms of thickness of the sample, volume fraction of the particles, and exposure time [10]. We can see that the scattering intensity measured in the present study is, as a whole, 10 times higher than that in the previous study because of the improvement of the monochromator and collimating optics. In particular, the intensity profile in the high-q range is clear, while that with the previous setting fluctuates considerably due to the readout noise of the CCD detector. We can also see the low-q limit of the profile is greatly improved. This is because the appropriate arrangement as Kratky camera consisting of S2, S3 and the photodiode results in reduced parasitic scattering.

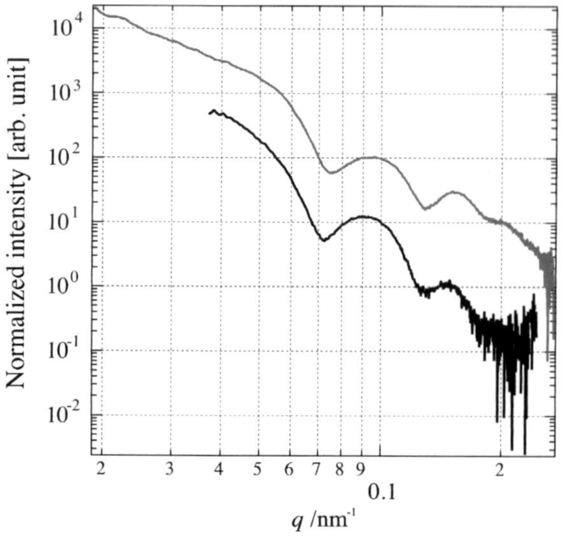

Figure 3. Scattering profiles obtained with two different settings. Red line is the profile obtained with the improved setting and black line is that of the previous setting. These profiles were normalized in terms of thickness of the sample, volume fraction of the particles, and exposure time. The energy of the X-ray beam was 2460 eV.

4. Summary

With the use of Si (111) channel-cut monochromator, the photon flux and the energy resolution increased by 10 times and 2.5 times, respectively, compared to those with the previous setting. In addition, collimating optics so-called Kratky camera was introduced to improve a low-q limit. Thus, SAXS intensity increased by about 10 times with the improved low-q limit. The meaningful high-q limit also extended to 0.02 nm^{-1} owing to the improved S/N ratio in SAXS intensity profile.

Acknowledgments

The authors are indebted to Drs. N. Ohta, Y. Suzuki, and K. Uesugi (JASRI) for their kind support on the experiment. The experiment was performed under the approval of JASRI program committee (Proposal No. 2010A1044).

References

[1] Lindner P and Zemb T (eds) 2002 *Neutron, X-ray and Light Scattering: Soft Condensed Matter* (Elsevier)
[2] Carugo K D, Helliwell J R, Stuhrmann H B, and Weiss M 2005 *J. Synchrotron Rad.* **12** 410.
[3] Handa M, Shinohara Y, Kishimoto H, Tamenori Y, and Amemiya Y 2010 *J. Phys. Conf. Ser.* 247 012006.
[4] Hettrick M C and Bowyer S 1983 *Appl. Opt.* **22** 3921.
[5] Ohashi H, Ishiguro E, Tamenori Y, Okumura H, Hiraya A, Yoshida H, Senba Y, Okada K, Saito N, Suzuki I, Ueda K, Ibuki T, Nagaoka S, Koyano I and Ishikawa T 2001 *Nucl. Instrum. Meth. A* **467-8** 533.
[6] Tamenori Y, Ohashi H, Ishiguro E and Ishikawa T 2002 *Rev. Sci. Instrum.* **73** 1588.
[7] Glatter O and Kratky O 1982 *Small Angle X-Ray Scattering* (Academic Press)
[8] Tamenori Y 2010 *J. Synchrotron Rad.* **17** 243.
[9] Matsushima T and Hashizume H 1983 *Handbook on Synchrotron Radiation* (Elsevier)
[10] Hubbell J H and Seltzer S M *http://www.nist.gov/pml/data/xraycoef/index.cfm*

Solvent Annealing Induced Perpendicular Orientation of Cylindrical Microdomains in Polystyrene-*b*-poly(4-hydroxyl styrene)/PEG Oligomer Blend Thin Film Made by Spin-coating from Selective Solvent

Taito Matsutani and Katsuhiro Yamamoto*

Department of Materials Science & Technology, Graduate School of Engineering, Nagoya Institute of Technology, Gokiso-cho, Showa-ku, Nagoya 466-8555, Japan

yamamoto.katsuhiro@nitech.ac.jp

Abstract. The microphase separated structure of PS-*b*-PHS/PEG blend thin film with thickness of 500 ~ 600 nm was investigated by grazing incidence small angle X-ray scattering. The thin film was obtained by two different solutions; one was THF which was common good solvent for all components of polymers used here. The other is toluene which was selective solvent for PS and poor-solvent for PHS and PEG. The equilibrium morphology of the block copolymer and blend sample was hexagonally packed cylinder in the bulk and thin film. The structure in the thin film obtained by spin cast from toluene solution was non-equilibrium. After THF vopar annealing of the thin film (cast from toluene), the highly ordered and perpendicular oriented cylindrical structure was obtained. Perpendicular orientation was failure when the thin film sample made by spin cast from THF solution and subsequent THF vapor annealing. The perpendicular nano-holes were fabricated after removing PEG oligomer by washing with water.

1. Introduction

Microphase separated structures of block copolymer thin films were extensively investigated by grazing incidence small angle X-ray scattering technique (GISAXS).[1-6] Controlling the morphology and particularly the orientation behaviour of phase-separated structures in both thin and thick films has received considerable attention because of their potential nanofabrication application. In the thin film, surface-polymer interaction, confinement effect, and the film thickness are really important factor to give a morphology that is different from those in the bulk. Some researchers have achieved perpendicular orientation of cylindrical morphology of block copolymer in thin film by controlling above factors. The study on orientation of microphase separated structure has been reported extensively. Solvent annealing [7-9], ion doping [5], surface modification of substrate (controlling interfacial energy) and thickness controlling related to the commensurability of domain spacing with film thickness [10,11] are effective ways to orient the microphase separated structures. In a thicker film, we reported spontaneous perpendicular orientation of cylindrical microdomains in a block copolymer normal cast film [12]. Selective solvent casting gave nonequilibrium morphology because one block fully shrank. Since vitrification of the shrunk chains occurred during evaporation of the solvent, the nonequilibrium morphology was finally arrested after completely dry. Appropriate annealing induced phase transition from nonequilibrium (sphere) to equilibrium structure (cylinder) and spontaneous orientation has been done. The direction of the cylinder orientation was concluded to be determined already by the direction of the solvent evaporation. This phenomenon can be referred to as an effect of a chemical potential gradient. In this paper, this method was applied to the thin film of ~ 500 nm in thickness (relatively thicker than usual film) and effect of initial morphology after spin-casting from common and selective solvents on equilibrium structure in the film will be reported.

2. Experimental

2.1. Materials

Toluene and tetrahydrofuran (THF) (Nacalai Tesque Inc., Extra Pure) were used as a solvent without further purification. In order to remove a stabilizer and inhibitors in monomers, styrene (Nacalai) and 4-*tert*-butoxy styrene (Wako Chemical, Inc.), granular activated alumina (Kanto Chemical Inc.) was added into the each monomer. Poly(ethyleneglycol) monometyl ether (PEG, M_n = 550) was purchased from (Aldrich Chemical Company). Styrene monomer was futher distilled under reduced pressure before polymerization. Silicon wafer (<0.02 Ω cm, 0.5mm thick) was obtained from the Nilaco Corporation. The wafer was washed with dilute sulphuric acid and rinsed with distilled water before use. The wafer was fully dried in air for 48 h.

2.2. Sample Preparation

A block copolymer polystyrene-*b*-poly(4-*tert*-butoxystyrene) (PS-*b*-PBxS) was synthesized by sequential anionic polymerization. ($M_n = 3.35 \times 10^4$ g/mol, $M_n/M_w = 1.10$, $f_{PBxS} = 19.5$ mol %). The *tert*-butoxy groups in the block copolymer were fully hydrolyzed by trifluoroacetic acid in toluene solution. The obtained PS-*b*-poly(4-hydroxyl styrene) (PHS) (ϕ_{PHS} = 19.9 vol%) was dissolved in toluene or THF solution. Subsequently, PEG was blended in the solutions until total volume fraction ϕ_{PS} of PS in polymers equaled to 30 vol%. The solutions were vigorously stirred at RT for 24 h. Total polymer concentration of the solution was 9.1 wt%. Thin films on the silicon wafer were obtained by spin coating of the solutions at room temperature without further thermal annealing. The spin rate was 3000rpm for 30 sec. The bulk samples (thick film, ~ 0.5 mm) were prepared by normal solvent casting from the solutions at room temperature and subsequently annealed at 423K for an hour. The thickness of thin films was analyzed by an atomic force microscopy. The sample thickness was more than 500 nm. The thin films were exposed to THF vapor (THF solvent annealing) for given time in an annealing chamber. After solvent annealing, the sample was taken out from the chamber and was set on the stage used for GISAXS measurement.

2.3. SAXS and GISAXS Measurements

The microphase separated structures was observed using small angle X-ray scattering (SAXS) and GISAXS. SAXS and GISAXS measurements were conducted at BL9C in Photon Factory of KEK, Tsukuba in Japan and BL40B2 and BL03XU in SPring-8, Hyogo in Japan. In BL9C, an imaging plate was used as a detector which was set at a position of 110 cm apart from sample position. In BL40B2, the charge-coupled device with an image intensifier (II-CCD, Hamamatsu Photonics Co., Ltd.) was used as a detector that was set at a position of 250 ~ 300 cm apart from sample position. The wavelength λ of X-rays was 0.15 (PF) and 0.10 (SPring-8) nm in the stations. Collagen (chicken tendon) was used as a standard specimen to calibrate the SAXS detectors. The grazing angle α_i of X-ray between X-ray and sample surface was in a range of 0.16 ~ 0.21° which is smaller than the critical angle (~0.22°) of the interface between the block copolymer and the substrate and bigger than that of the block copolymer surface (~0.15°).

3. Result and Discussion

3.1. Morphology in bulk and solutions.

The morphology of the synthesized PS-*b*-PHS and PS-*b*-PHS/PEG Blend was hexagonally packed cylindrical structure as shown in Figure 1a. These films were obtained after casting from THF solution and subsequently annealing at 423K. Even if PEG oligomer was added in the sample, the cylindrical morphology was maintained except for domain spacing. The domain spacing increased due to swelling of PHS cylinder domains by incorporation of PEG into them. How about the morphology in the solutions? THF is a common good solvent for all of components. Toluene is good solvent for PS but poor solvent for PHS and PEG. In a common solution (THF), no structure was observed at a concentration of 10wt% as shown in Figure 1b. The common solvent diluted segregation power χN by factor ϕ_S (χ is the Flory-Huggins interaction parametner, N is the total number of degree of polymerization, ϕ_S is the solution concentration). A decrease in the $\phi_S\chi N$ put the system into disordered phase. On the other hand, in a selective solvent (toluene), the segregation power expected to be increased and the chains of PHS and PEG shrunk in toluene solution. These situations gave the much different morphology in comparison with that in THF solution. Figure 1b shows the microphase separated structure which was not completely assigned but probably spherical morphology in toluene. The thin films were obtained by spin-coating from these solutions.

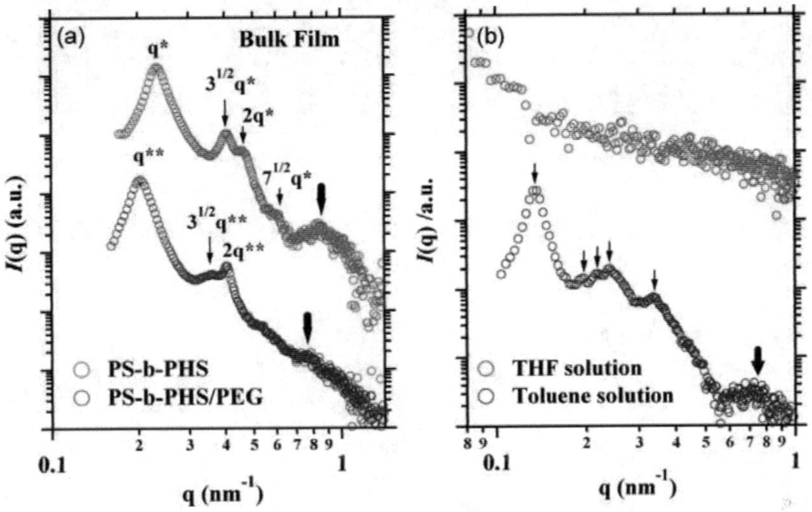

Figure 1. SAXS profiles of (a) PS-*b*-PHS and PS-*b*-PHS/PEG blend bulk films and (b) PS-b-PHS/PEG in THF (upper) and toluene (bottom) solutions. Thin and thick arrows indicated lattice and form factors, respectively.

3.2. Morphology in thin films

After spin coating without further thermal annealing, GISAXS measurement was conducted to reveal the structure in the thin films. As for cast from THF solution, there was no significant pattern in Figure 2a, that is, no clear structure in the film. On the other hand, longitudinally long elliptic arc pattern appeared in the film made by spin-coated from toluene solution as shown in Figure 3a. This pattern was scattering from the microphase-separated spherical structure as appeared in the solution except for lattice disorder. After the thin films were exposed to THF vapor (solvent annealing), specific patterns were found for both films. In the case of cast from THF solution, new scattering spots were clear after annealing for 4 h (Figure 2b). In contrast, diffractions of *in-plane* (q_y) direction and vertically long streak-like pattern were observed in the films cast from toluene solution after annealing for only 5 min (Figure 3b). Much long annealing achieved an effect on ordering of the structure for both samples. Hexagonal pattern was obtained in the thin film from THF solution after annealing for 24 h (Figure 2c). This pattern was given by the hexagonally arrayed cylinders oriented parallel to the substrate. It is common case that the parallel orientation appears in the block copolymer with cylindrical morphology.

On the contrary, significantly clear higher-ordered diffraction peaks developed in only *in-plane* direction in the film obtained from THF solution after annealing for 4 h as indicated in Figure 3c. This feature in GISAXS pattern showed that the cylinders in the thin film were aligned perpendicularly to the substrate. The degree of the orientation was fairly high because of no diffraction spots in out-of-plane (q_z-direction). One dimensional scattering profiles of *in-plane* direction for the toluene cast thin film was shown in Figure 4. Before annealing, the broad first order peaks were seen at q_y of 0.27 nm^{-1} and weak broad peak was observed at q_y of 0.7 nm^{-1} which was probably from particle scattering of spheres and coincided with the position in the SAXS profile in the toluene solution. THF common solvent annealing sharpened scattering peaks and higher ordered peaks were clearly observed at a relative peak positions of $1 : 3^{1/2} : 2 : 7^{1/2} : 3$. That is, the cylinder domains were packed hexagonally in the film.

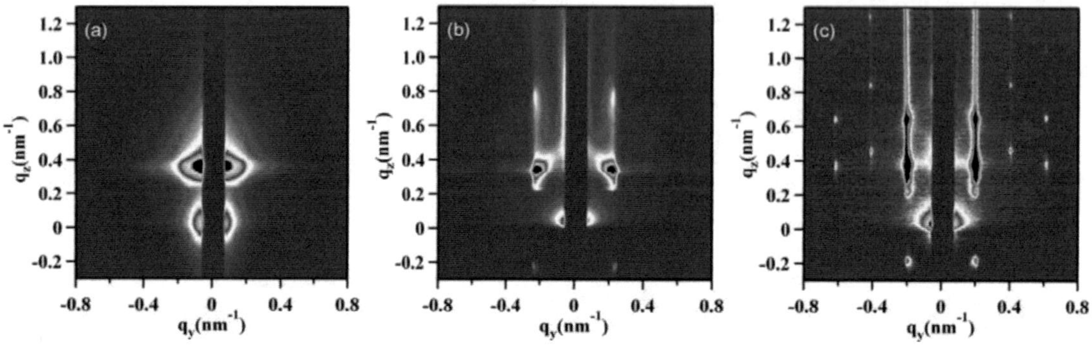

Figure 2. GISAXS patterns of the thin films obtained by spin-coating from THF solution: (a) as spun, THF vapor annealing (b) for 5 min and (c) for 4 h.

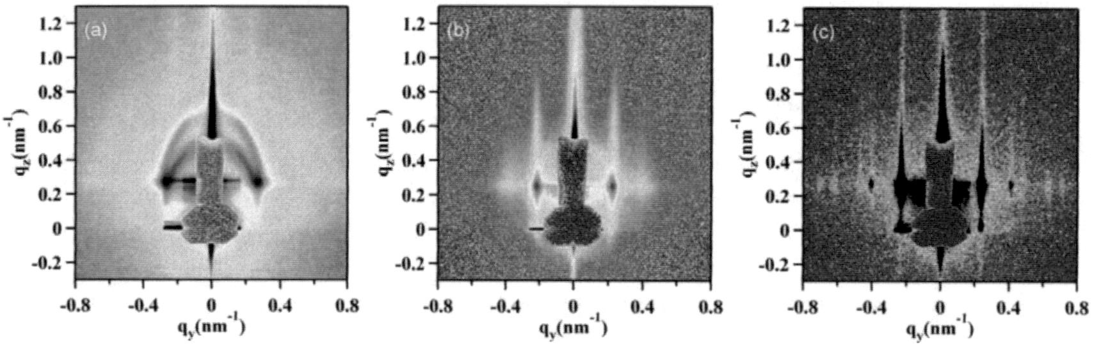

Figure 3. GISAXS patterns of the thin films obtained by spin-coating from toluene solution: (a) as spun, THF vapor annealing (b) for 5 min and (c) for 4 h.

It is important to know the mechanism of the perpendicular orientation. This is considered to be analogous behavior to our recent work about spontaneous perpendicular orientation of cylindrical microdomains in a block copolymer [12]. Selective solvent casting gave non-equilibrium morphology in the film. In this study, the non-equilibrium morphology (sphere) was finally arrested after completely dry. Phase transition from non-equilibrium (sphere) to equilibrium structure (cylinder) by annealing and spontaneous orientation has been done. The direction of the cylinder orientation was considered to be determined already by the direction of the evaporation of selective solvent. In the case of casting from THF solution, the disordered state was first set as spun film. The subsequently annealing induced phase separation and ordering of it. However, the direction of structure development was not controlled.

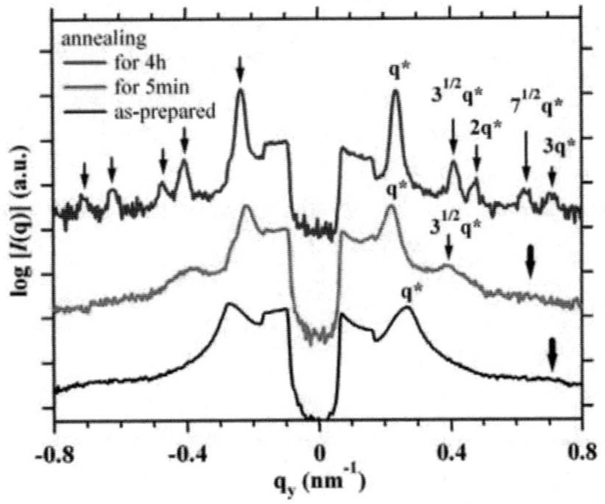

Figure 4. *In-plane* 1D-SAXS profiles at q_z = ~ 0.25° obtained by spin-coating from toluene solution: As-prepared (bottom), THF vapor annealing (middle) for 5 min and (top) for 4 h. Thin and thick arrows indicated lattice and form factors, respectively.

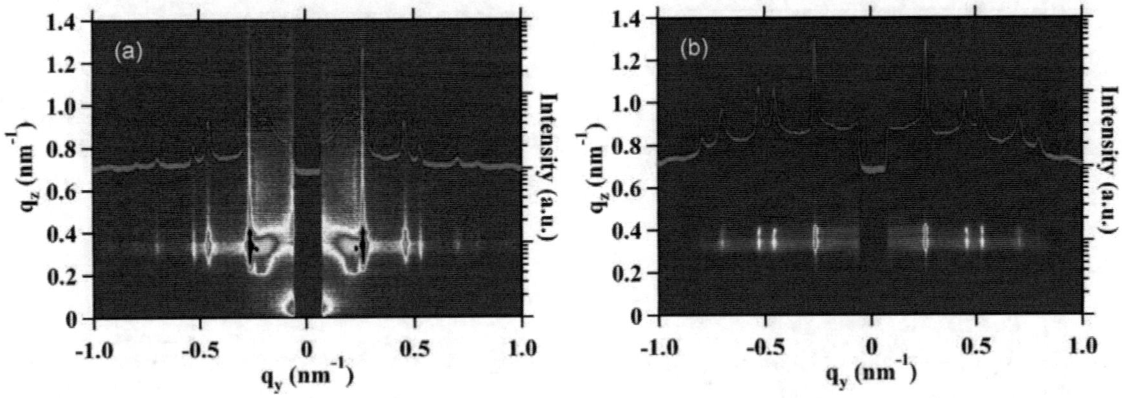

Figure 5. GISAXS profiles of PS-*b*-PHS/PEG thin film obtained by spin-cast from toluene solution. The sample was annealed with THF vopar. The GISAXS measurement (α_i = 0.18°) was conducted (a) before and (b) after water washing with exposure (data accumulation) time of X-ray of 250 and 10 ms, respectively. One-dimensional profiles (red lines, right axis) in the Figures indicated *in-plane* scatterings at q_z ~ 0.35.

Finally, we tried to remove the blended PEG oligomer from the film with highly perpendicular oriented cylinders by soaking thin film on the silicon wafer into distilled water. The intensity of GISAXS increased by a factor of approximately 20 ~ 30 after washing without any change in peak positions as shown in Figure 5. The perpendicular oriented cylindrical structure was maintained even after washing with water. The increase in the GISAXS intensity was ascribed to removing PEG from cylinders and nano-holes were created in the film. The generation of nano-holes enhanced the contrast between PS matrix and the cylinders. One-dimensional profiles in the Figure (red lines) indicated *in-plane* scatterings at q_z ~ 0.35. These profiles are slightly different except for peak positions. The relative intensity of the higher order peaks to the first order peak increased after removing PEG. The change in the intensity ratio is ascribed to the electron-density profile. In the thin film, a core-shell like cylinder (PHS cylinder including air near the core) can be generated after removing PEG oligomer. The values of the electron density ρ_e for each polymer were calculated to be 0.565 for PS and 0.629 mol/cm^3 for PHS. The PHS chains can be located near the wall of the cylinder and the concentration of

PHS chains can be imagined to gradually decrease toward the core of the cylinder. In this assumption, the values of the ρ_e near the wall of the cylinder and the core of the cylinder are higher and lower than the ρ_e of PS matrix. In such a case, intensity of the first order peak diminishes relative to that of the higher order peaks. Therefore, the nano-hole structure was predicted to be core-shell type cylinder. Here, let us think the increment of the GISAXS intensity. The electron density difference $\Delta\rho_e$' between PS matrix and PHS/PEG cylinder as assuming that PHS and PEG fully mix was calculated to be 0.051 mol/cm^3 where the ρ_e for PEG 0.594 mol/cm^3, respectively, and the volume fraction of PEG in the cylinder is 0.43. After removing PEG, the $\Delta\rho_e$" between PS matrix and PHS/air cylinder was estimated roughly at 0.25 mol/cm^3 (PHS/air domain $\rho_e \sim$ 0.315 mol/cm^3). SAXS intensity is proportional to the square of $\Delta\rho_e$. The value of $(\Delta\rho_e")^2$ was 24 times as large as that of $(\Delta\rho_e')^2$. This value is coincidence with the enhancement of GISAXS intensity. Additionally, we took into consideration the depth of the nano-hole. The *in-plane* profile (not shown) at α_i of 0.14° which was lower than the critical angle of the polymer-air interface was almost identical with the profile in Figure 5b. Namely, there was no significant structural difference between the entire film and the vicinity of the surface. This means that the depth of the nano-hole attained the film thickness. However, more quantitative and accurate calculation are necessity to reveal the more detailed structure in the thin film.

Acknowledgment
The SAXS and GISAXS measurements were performed at he Photon Factory of High Energy Accelerator Research Organization (Approval 2008G027, 2008G525, 2010G028) and at SPring-8 (Approval 2009B1103, 2010A1180, 2010A7225). This research was supported by Ministry of Education, Science, Sports, and Culture, Grant-in-Aid for Young Scientists (B) (20750176, 2008). Authors thank to Prof. Shinichi Sakurai (Kyoto Institute of Technology), Prof. Shusaku Nagano (Nagoya University) and Dr. Mitsuo Hara (Nagoya University) for their kind help and fruitful discussion.

References

[1] Lee B, Park I, Yoon J, Park S, Kim J, Kim K-W, Chang T and Ree M 2005 *Macromolecules* **38** 4311
[2] Jin S, Yoon J, Heo K, Park H.-W, Kim J, Kim K-W, Shin T J, Chang T and Ree M 1997 *J.Appl. Cryst.* **40** 950
[3] Stein G E, Kramer E J, Li X and Wang J 2007 *Macromolecules* **40** 2453
[4] Yokoyama H, Dutriez C, Li L, Nemoto T, Sugiyama K, Sasaki S, Masunaga H, Takata M, and Okuda H 2007 *J. Chem. Phys.* **127** 014904
[5] Wang J Y, Chen W, Roy C, Sievert J D and Russell T 2008 *Macromolecules* **41** 963
[6] Busch P, Posselt D, Smilgies D –M, Rauscher M and Papadakis C M 2007 *Macromolecules* **40** 630
[7] Bang J, Kim B J, Stein G E, Russell T, Li X, Wang J, Kramer E J and Hawker C J 2007 *Macromolecules* **40** 7019
[8] Xu T, Goldbach J T, Misner M J, Kim S, Gibaud A, Gang O, Ocko B, Guarini K W, Black C T, Hawker C J and Russell T 2004 *Macromolecules* **37** 2972
[9] Cavicchi K and Russell T 2007 *Macromolecules* **40** 1181
[10] Potemkin I I 2004 *Macromolecules* **37** 3505
[11] Horvat A, Knoll A, Krausch G, Tsarkova L, Lykhova K S, Sevink G J A, Zvelindovsky A V and Magerle R 2007 *Macromolecules* **40** 6930
[12] Sakurai S, Bando H, Yoshida H, Fukuoka R, Mouri M, Yamamoto K and Okamoto S 2009 *Macromolecules* **42** 2115
[13] Hashimoto T, Kawamura T, Harada M and Tanaka H 1994 *Macromolecules* **27** 3063
[14] Bosch P, Rauscher M, Smilgies D-M, Posselt D, Papadakis C M, 2006 *J. Appl. Crystallogr.* **39** 433.

Journal of Physics: Conference Series **272** (2011) 012016 doi:10.1088/1742-6596/272/1/012016

Static and Dynamic Scattering from Polysulfobetaine Immobilized on Silica Nanoparticle in Ionic Liquid

Moriya Kikuchi[1], Yuki Terayama[2], Taiki Hoshino[1], Motoyasu Kobayashi[1], Hiroki Ogawa[3], Hiroyasu Masunaga[3], and Atsushi Takahara[1,2,4]

[1]JST, ERATO Takahara Soft Interfaces Project, 744 Motooka, Nishi-ku, Fukuoka 819-0385, Japan, [2]Graduate School of Engineering, Kyushu University, 744 Motooka, Nishi-ku, Fukuoka 819-0385, Japan, [3]Japan Synchrotron Radiation Reserch Institute / Spring-8, 1-1-1, Kouto, sayo-cho, Sayo-gun, Hyogo 679-5198, Japan, and [4]Institute for Materials Chemistry and Engineering, Kyushu University, 744 Motooka, Nishi-ku, Fukuoka 819-0385, Japan

E-mail: takahara@cstf.kyushu-u.ac.jp

Abstract. Monodisperse silica nanoparticles were immobilized with poly[3-(*N*-2-methacryloyloxyethyl-*N*,*N*-dimethyl)ammonatopropanesulfonate)] (SiNP-PMAPS) by surface-initiated atom transfer radical polymerization. SiNP-PMAPS was characterized in 1-hexyl-3-methylimidazolium chloride (HMImCl) solution by dynamic light scattering (DLS) and synchrotron small angle X-ray scattering (SAXS) measurements. The relaxation spectrum of DLS and SAXS profile for SiNP-PMAPS solutions showed a unimodal distribution attributed to the homogeneous dispersion of SiNP-PMAPS in the HMImCl without any aggregations. The scattering factor from the SiNP-PMAPS in HMImCl was described by the core-shell model considering interacting excluded-volume chains and assuming Schulz distributed core with the thickness of polymer layer of 56 nm.

1. Introduction

Sulfobetains are special classes of zwitterions having ammonium cation and sulfonate anion isolated with some methylene spacer, such as propyl and butyl group. The poly[3-(*N*-2-methacryloyloxyethyl-*N*,*N*-dimethyl)ammonatopropanesulfonate)] (PMAPS) exhibit a unique aqueous solution behavior in contrast with the conventional polyelectrolytes. For instance, the PMAPS are insoluble in pure water and can dissolve in above 60 mM NaCl aqueous solution at 30 °C [1]. The chain dimension of the PMAPS is dependent on the attractive polyampholyte effect (electrostatic attractive forces) between ammonium cations and sulfonate anions of the side chains, and also on the non-ionic excluded-volume effect between segments. The ionic liquids (ILs) are interesting dispersion media for colloidal particles as well as aqueous solution. The ILs as novel classes of solvent are composed entirely of ions and are fluid at ambient conditions. The ILs exhibit the unique physicochemical properties, such as negligible vapour pressure, nonflammability, and distinguished thermal and chemical stabilities. However, the chain dimension of the PMAPS brush state on silica nanoparicle surface in ionic liquid is still unclear.

In this work, we carried out dynamic light scattering (DLS) and synchrotron small-angle X-ray scattering (SAXS) measurements of polymer-grafted silica nanoparticle (SiNP) in 1-hexyl-3-

methylimidazolium chloride (HMImCl) with high ionic strength (5.19 M) and high viscosity (155 g cm^{-1} s^{-1}) to investigate the chain dimension of PMAPS chains.

2. Experiments

SiNP (diameter of 108 nm; Nissan Chemicals) was dispersed in a mixture of ethanol/water/ammonia containing (2-bromo-2-methyl)propionyloxyhexyltriethoxysilane (BHE) to immobilize alkylbromide on the surface [2]. The surface-initiated atom transfer radical polymerization of 3-(*N*-2-methacryloyloxyethyl-*N*,*N*-dimethyl)ammonatopropanesulfonate) (MAPS) mediated by a copper complex was carried out with the initiator-immobilized SiNP (SiNP-BHE) in 2,2,2-trifluoroethanol (TFE) containing a small amount of 1-hexyl-3-methylimidazolium chloride (HMImCl) at 60 °C [3]. The number-average molecular weight (M_n) and polydispersity index (M_w/M_n) of the cleaved graft PMAPS from the SiNP surface with HF aqueous solution were determined to be $M_n = 2.06 \times 10^5$ g mol^{-1} and $M_w/M_n = 1.21$ by size exclusion chromatography calibrated with a series of PMAPSs as a standard. The graft density of PMAS brush on SiNP surface was estimated to be 0.071 chains nm^{-2} using values of weight loss measured by thermogravimetric analysis of SiNP-PMAPS and M_n of PMAPS. The average number of graft PMAPS chain per SiNP was estimated to be 2550 by the graft density and the SiNP surface area.

DLS measurements of SiNP-PMAPS in HMImCl at 25 °C were carried out at scattering angle (θ) range from 30 to 150 ° using a goniometer system purchased from ALV (Langen, Germany) employing a HeNe laser (wavelength λ = 632.8 nm) with a power of 22 mW. The autocorrelation function is obtained by pseudo-cross-correlation of the signals from two photomultipliers to suppress noise. The DLS measured samples were four concentrations (C_p = 0.527 - 2.07 mg cm^{-3}) of SiNP-PMAPS in HMImCl contained in 10 mm diameter glass cells. The scattering vector (q) was defined as $4\pi n_0 \sin(\theta/2)/\lambda$, where n_0 is the refractive index of HMImCl.

SAXS meaurements of SiNP-PMAPS in HMImCl at room temperature was carried out at the BL03XU beamline of SPring-8 using an incident X-ray with a wavelength λ = 0.1 nm and sample-to-detector distance of 6103 mm. The scattered intensity was detected by an imaging plate with 3000 × 3000 pixels over a scattering vector range from 0.013 to 1.5 nm^{-1}. The scattering vector (k) as defined by $4\pi\sin(\theta)/\lambda$ was calibrated by the average of 8 peaks of dry collagen. The SAXS measured samples were four concentrations (C_p = 1.03 - 1.81 mg cm^{-3}) of SiNP-PMAPS in HMImCl contained in 2 mm diameter quartz capillaries.

3. Results and Discussion

The typical autocorrelation function ($g^{(2)}(t)$-1) and relaxation spectrum ($A(\tau)$) measured by DLS for the SiNP-PMAPS in HMImCl are shown in Figure 1 (a). The concentration of the SiNP-PMAPS and the scattering angle were 2.07 mg cm^{-3} and 90 °, respectively. On the ordinate, the $A(\tau)$ at relaxation time (τ) was calculated from the intensity of $g^{(2)}(t)$-1 at time (t) by CONTIN analysis. The $A(\tau)$ was observed as a single and sharp peak. The $A(\tau)$ of all concentrations and scattering angles for the SiNP-PMAPS solutions showed a unimodal distribution and sharp peak indicating the homogeneous dispersion of SiNP-PMAPS in HMImCl without any aggregations.

The mutual diffusion coefficient (D_m) and hydrodynamic radius (R_H) calculated from $A(t)$ are expressed by the following equations:

$$D_m = \lim_{q\to 0}\left(\Gamma q^{-2}\right) = \lim_{q\to 0}\frac{\sum_{i\in \text{peak}} A(\tau_i)\tau_i^{-1}}{q^2 \sum_{i\in \text{peak}} A(\tau_i)} \quad \text{where} \sum_{i=0}^{\infty} A(\tau_i) = 1 \tag{1}$$

$$R_H = \frac{k_B T}{6\pi\eta_0 D_0} \tag{2}$$

where Γ, k_B, T, η_0, and D_0 are the first cumulant, the Boltzman constant, the absolute temperature, viscosity of the solvent, and the diffusion coefficient at an infinite dilution, respectively.

Figure 1 (b) shows concentration dependence of the D_m at the zero scattering vector for the SiNP-PMAPS in HMImCl at 25 °C. By extrapolating to $C_p = 0$, the D_0 and R_H for the SiNP-PMAPS in HMImCl at an infinite dilution were estimated to be 1.57×10^{-12} cm s^{-1} and 89.8 nm, respectively.

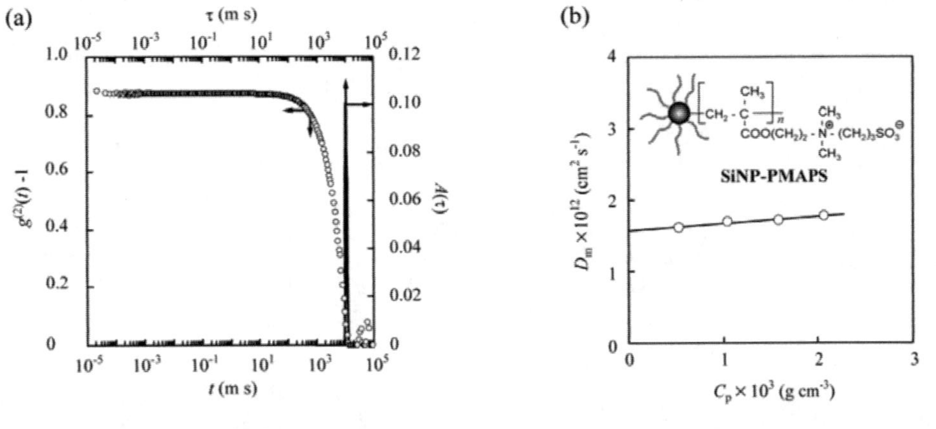

Figure 1. (a) Typical autocorrelation function (○) obtained by DLS for SiNP-PMAPS in HMImCl (θ = 90 °) and relaxation spectrum (line) calculated from the autocorrelation function. The concentration of Si-PMAPS in HMImCl and the scattering angle are C_p = 2.07 mg cm^{-3} and 90 °. (b) Concentration dependence of the measured D_m for SiNP-PMAPS in HMImCl at 25 °C.

Figure 2 (a) shows the scattering form factor ($\Delta I(k)/\Delta I(0)$) from the SAXS profile for the SiNP-BHE in TFE (1) with C = 1.38 mg cm^{-3}. Similar four minima were observed in the SiNP-BHE form factor. The theoretical curve was also plotted in Figure 2 (a) by assuming Schulz distributed spheres [4] with a radius of R = 53.5 nm and a polydispersity of p = 0.06. The experimental SAXS profile for the SiNP-BHE in TFE was quantitatively described by sphere model with assuming Schulz distribution.

The SAXS profile for the SiNP-PMAPS in HMImCl (2) with C_p = 1.81 mg cm^{-3} showed smooth minima in the SiNP-PMAPS form factor which was slightly shifted to a low k region in comparison with that of the SiNP-BHE form factor. The SiNP-PMAPS in HMImCl is expected to form core-shell structure, as shown in the inset of Figure 2 (a). Here we estimated radius of SiNP-PMAPS from the SAXS profile using the core-shell model analysis considering interacting excluded-volume chains [5]. The scattering form factor for core-shell model containing four terms is expressed by

$$P(k) = \begin{Bmatrix} N^2 \Delta Z_{\mathrm{e,core}}^2 P_{\mathrm{core}}(k) + N \Delta Z_{\mathrm{e,chain}}^2 P_{\mathrm{PWC}}(k) \\ + 2N \Delta Z_{\mathrm{e,core}} \Delta Z_{\mathrm{e,chain}} S_{\mathrm{core-chain}}(k) \\ + N[N - P_{\mathrm{PWC}}(0)] \Delta Z_{\mathrm{e,chain}}^2 S_{\mathrm{chain-chain}}(k) \end{Bmatrix} N^{-2} \left(\Delta Z_{\mathrm{e,core}} + \Delta Z_{\mathrm{e,chain}} \right)^{-2} \quad (3)$$

where the N is the number of graft polymer chain, ΔZ_e is the excess electron density, $P_{core}(k)$ is scattering function for rigid sphere with assuming Schulz distribution, $P_{PWC}(k)$ is scattering function for infinitely thin perturbed wormlike chain [6], $S_{core\text{-}chain}(k)$ is structure factor between the core and chains, and $S_{core\text{-}chain}(k)$ is structure factor between different chains.

The theoretical curve calculated by equation 3 with the radius of core R_c = 53.5 nm, the polydispersity of core p = 0.06, the radius of polymer layer of s = 28 nm is shown in Figure 2 (a). The experimental SAXS profile for SiNP-PMAPS in HMImCl was close to the calculated values based on from core-shell model and $P(k)$ function, which related with $\Delta I(k)/\Delta I(0)$.

Figure 2 (b) shows the radial profiles for SiNP-PMAPS in HMImCl estimated from the SAXS profile and the core-shell model with interacting excluded-volume chains. The M_w-dependence of z-average mean-square radius of gyration ($\langle S^2 \rangle_z$) for the free linear PMAPS in 74 mM aqueous NaCl solution at 25 °C as a theta state is given by the following equation:

$$\langle S^2 \rangle_z^{1/2} = 0.017 M_w^{0.5} \quad (4)$$

The value of polymer layer ($2s$ = 56 nm) was lager than the $2\langle S^2\rangle_z^{1/2}$ for free linear PMPAS chain with same molecular weight and polydispersity in theta state (noninteracting) calculated from equation (4), as shown in Figure 2 (b). The ratio of radius of gyration (56 nm) measured by SAXS to hydrodynamic radius (36.3 nm) measured by DLS for PMAPS layer of SiNP-PMAPS in HMImCl was estimated to be 1.54. Thus, chain conformation of PMAPS brush on SiNP surface is behaved as Gaussian coil.

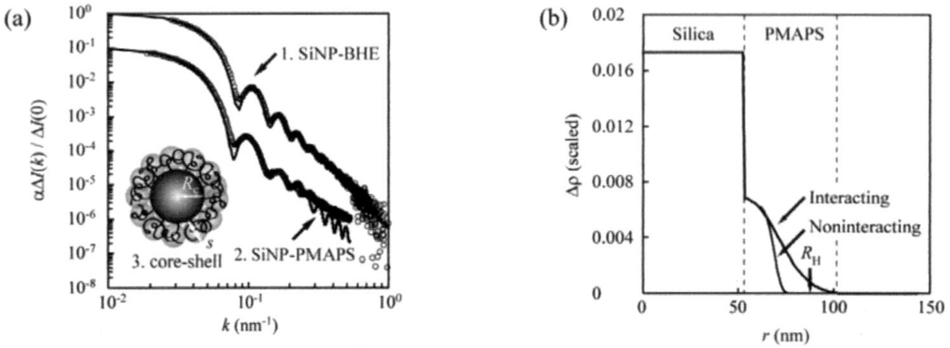

Figure 2. (a) SAXS profiles of SiNP-BHE in TFE (1) and SiNP-PMAPS in HMImCl (2), together with the schematic cross-section drawing of a core-shell model (3). The solid curves represent the theoretical values for the sphere model with assuming Schulz distribution and the core-shell model taking into interacting excluded-volume chains and assuming Schulz distribution core. (b) Radial electron density profiles for SiNP-PMAPS in HMImCl estimated from SAXS profile and core-shell model.

4. Conclusion

The relaxation spectrum of all concentrations and scattering angles for SiNP-PMAPS solutions by DLS showed a unimodal distribution and sharp peak attributed to the homogeneous dispersion of SiNP-PMAPS in the HMImCl without any aggregations. The scattering factor from the SiNP-PMAPS in HMImCl is described by the core-shell model considering interacting excluded-volume chains and assuming Schulz distributed core with the thickness of polymer layer of 56 nm. The thickness of polymer layer estimated from SAXS profile is lager than the radius of gyration for the free PMAPS chain with same molecular weight in theta state. From the ratio of radius of gyration to hydrodynamic radius for PMAPS layer of SiNP-PMAPS in HMImCl, chain conformation of PMAPS brush on silica surface is behaved as Gaussian coil.

Acknowledgement

Dr. Yasushi Okamoto of Denso Corporation is acknowledged for their help on SAXS measurements. The synchrotron radiation experiments were performed at BL03XU in the Spring-8 with the approval of the Japan Synchrotron Radiation Research Institute (JASRI) (Proposal No. 2010A7220).

References

[1] Kato T and Takahashi A 1996 *Ber. Bunsenges. Phys. Chem.* **100** 784
[2] Ohno K, Morinaga T, Koh K, Tsujii Y and Fukuda T 2005 *Macromolecules* **38** 2137
[3] Terayama Y, Kikuchi M, Kobayashi M and Takahara A 2010 *Polym. Prep. Japan* **59** 723.
[4] Aragón S R and Pecora R 1976 *J. Chem. Phys.* **64** 2395
[5] Pedersen J S, Svaneborg C, Almdal K, Hamley I W and Young R N 2003 *Macromolecules* **36** 416
[6] Nakamura Y and Norisuye T 2004 *J. Polym. Sci. B: Polym. Phys.* **42** 1398

Journal of Physics: Conference Series **272** (2011) 012017 doi:10.1088/1742-6596/272/1/012017

Advanced Neutron Reflectometer for Investigation on Dynamic/Static Structures of Soft-Interfaces in J-PARC

Koji Mitamura[1),2)], Norifumi L. Yamada[3)], Hidenori Sagehashi[3)], Hideki Seto[3)], Naoya Torikai[4)], Tsukasa Sugita[5)], Michihiro Furusaka[5)] and Atsushi Takahara[1),2)],
[1)]JST, ERATO Takahara Soft Interfaces Project, [2)]Institute for Materials Chemistry and Engineering, Kyushu University,[3)] High Energy Accelerator Research Organization (KEK), [4)]Graduate School of Engineering, Mie University, [5)] Graduate School of Engineering, Hokkaido University, Japan

E-mail: takahara@cstf.kyushu-u.ac.jp

Abstract. A novel neutron reflectometer with horizontal geometry will be established at BL16 in Materials and Life Science Experimental Facility (MLF) of Japan Proton Accelerator Research Complex (J-PARC) as a successor of a reflectometer ARISA-II. ARISA-II corresponding to a single neutron beam downward at 2.22 deg has achieved off-specular and time-resolved reflectivity measurements. The novel reflectometer is designed so as to receive two tilted neutron beams (2.22 & 5.71 deg), which gives us an opportunity in investigation on a free liquid surface. The reflectometer can provide a micro-sized beam by slit collimation and obtain a fair reflectivity with small sample area. Also, T0 chopper and neutron focusing mirror are newly introduced. The T0 chopper can suppress the background due to fast neutrons. The focusing mirror produces further reduction of measurement time not only for specular reflection by focusing neutrons on a sample, but also grazing incidence small-angle neutron scattering (GISANS) measurements by focusing on a detector.

1. Introduction

Neutron reflectometry is a powerful method for investigations on interfaces composed of soft materials so-called Soft-Interfaces, since neutrons can distinguish soft materials with and without deuteration [1]. The method allows us to measure the reflection from buried interfaces such liquid/solid, because of high transmittance of neutrons inside materials. Also, white pulsed neutron can cover wide range of scattering vector (q) at the same time, owing to time-of-flight method, which is applicable in time-resolved measurements to measure the dynamics of Soft-Interfaces [2].

ARISA-II is a horizontal-type neutron reflectometer installed at BL16 in Materials and Life Science Experimental Facility (MLF), J-PARC, Japan. For free interfaces, such as water surface, two downward beam lines with 2.22 and 5.71 deg relative to horizontal are introduced. ARISA-II was open for users from 2009, and made the fruits as a scientific publication [3, 4]. However, ARISA-II can receive only a 2.22 deg beam line because this reflectometer is reutilizing old ARISA reflectometer at KENS facility in KEK, Japan, [5] which does not have the enough strokes to change the vertical positions of sample and detector for both beam lines. This did not meet the demand on measurements with high q range for free liquid surfaces. Additionally, the minimum beam size of ARISA was relatively large (~0.1 mm), in which the footprint on sample is over 30 mm at incident angle of 0.3 deg with q resolution of 3 %. In the point of efficiency on sample preparations, further reduction of sample area, hopefully less than 10 mm × 10 mm, is preferred.

We are developing an advanced neutron reflectometer with horizontal geometry as a successor of ARISA-II in order to overcome the disadvantages of ARISA-II and realize the reflectometer with higher performance. In this paper, we describe about the present state of ARISA-II and discuss on the specification of the novel neutron reflectometer with introduction of new attempts for further upgrades.

2. Instrumental Outline of ARISA-II

Figure 1 shows the schematic drawing of ARISA-II. Pulsed neutron beam at 25 Hz came through a guide tube with $3Q_c$ super mirror. The beam passed through a disk chopper to determine wavelength band of incident neutrons as 2.0 – 8.8 Å. After the disk chopper, iron collimator was installed for reduction of unnecessary fast neutrons which give rise to an increase of background. Before a sample, the beam was finely collimated by a pair of silts (S1 and S2) with sintered B_4C blades. The samples were placed on a stage with horizontal geometry. The reflected neutrons went through the third slit and the reflection angle and time-of-flight of each neutron were counted by a 2-dimensional scintillation detector with ^{6}LiF/ZnS (OHYO KOKEN, Japan) and position-sensitive photomultiplier tube (HAMAMATSHU, Japan). The beam divergence, sample position, incident and reflection angles were precisely controlled by a computer to keep an angle resolution. The reflection intensity was normalized by incident beam intensity and converted to the reflectivity depending on scattering vector $\boldsymbol{q}$.

3. Results of Soft-Interfaces.

Figure 2 represents the time-resolved specular neutron reflectivity (NR) profiles of a poly(3-(N-2-methacryloyloxyethyl-N,N-dimethyl)ammonatopropanesulfonate) (PMAPS) brush on a quartz substrate in contact with D_2O. The PMAPS brush was synthesized by atomic transfer radical polymerization [6]. The first reflectivity curve was collected in 30 min after the immersion in a D_2O medium, and each reflectivity curve was obtained at scan duration of 300 seconds. The figure shows that weak fringes resulted from a brush structure in D_2O [4] were successfully obtained and the brush structure became stable within 30 min after the immersion of D_2O. The profiles indicate that a fair reflectivity profile with a statistical error of reflectivity ($\Delta R/R$) < 15 % can be obtained from liquid/solid interfaces in minute order within q_z = 0.03 Å^{-1}, where structural change of a film with thickness over 200 Å can be detectable. It should be noted that this experiment was performed at proton beam power of 120 kW (neutron flux evaluated by Monte Carlo simulation is 2.4×10^5 neutrons/s in this experiment). Since the J-PARC accelerator is planning to be 1MW, the time resolved measurements in about 40 sec as a scan slicing will be possible in the future. The measurements are applicable for interfacial dynamics of soft materials with dimension over 200 Å taking place in minute order.

Since ARISA-II utilized a 2-dimensional scintillation detector with the 100mm effective area in a diameter and 1 mm spatial resolution, off-specular reflectivity besides the specular reflection was

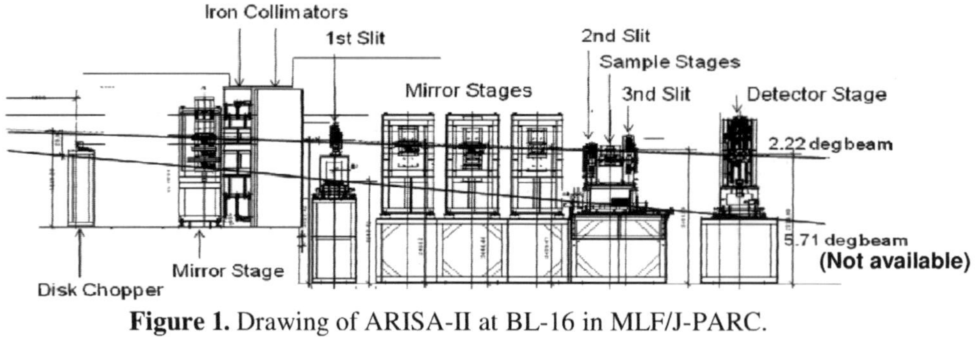

Figure 1. Drawing of ARISA-II at BL-16 in MLF/J-PARC.

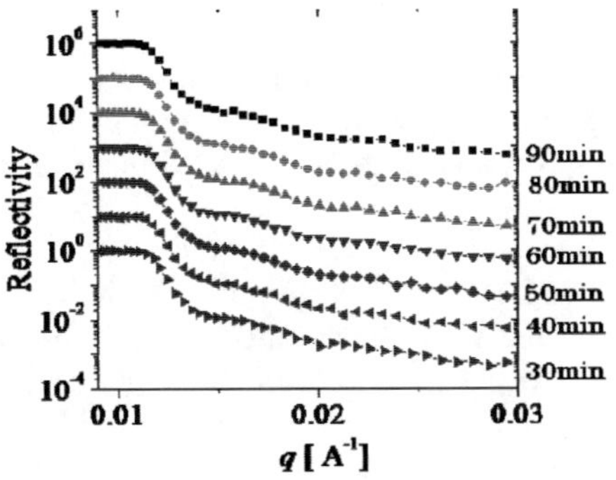

Figure 2. Time evolution of NR profiles obtained from PMAPS brush in contact with a D_2O medium.

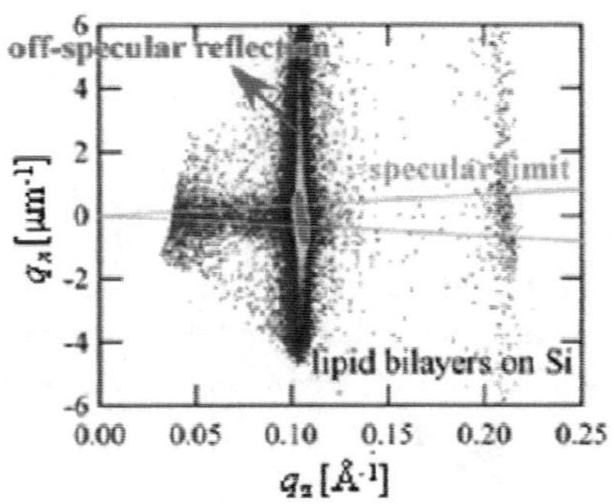

Figure 3. Two-dimensional NR profile from phospholipid bilayers along with q_z and q_x.

measured. Figure 3 shows NR profile obtained from a stacking of dipalmitoylphosphatidylcholine (DPPC) bilayers on a Si wafer with ARISA-II. As seen in the figure, the Bragg peaks resulted from lamellar structure of the DPPC bilayers were clearly observed at q_z= 0.11 and 0.22 Å $^{-1}$. Besides the Bragg peaks, the clear streaks in q_x direction were observed, where q_x is a momentum in direction of beam's travel among axes parallel to the film plane. The streaks of off-specular reflection was extended over q_x = 6 μm^{-1}. This indicates the presence of in-plane structures with long-length correlation, that is, the surface fluctuation of the multilayer as shown in a previous study [7]. In principle, grazing incidence small-angle neutron scattering (GISANS) measurements can be also realized by collimating the neutron beam in vertical direction.

4. Instrumental upgrades

For further upgrades, we are planning to replace old ARISA system as well as install new components. As presented in the proceeding result section, ARISA-II showed a good performance on specular and off-specular reflectivity measurement with the 2.22 degree beam line. To utilize the 5.71 degree beam line, a new reflectometer system with slits, sample and detector stages will be replaced to adjust their vertical positions with longer strokes. As simple calculation, an incident neutron with λ = 2.0 Å at 5.71 deg obtains q value 0.56 Å $^{-1}$ in a specular condition. The use of this beam will enable us to perform high q scans, which are of great use in the measurements on a liquid surface, especially in structural estimation of a monolayer or thin film with a few nanometers in thickness on a liquid surface (e.g. Langmuir or Gibbs film).

As well as the expansion of the vertical strokes, the fine silt collimation will be installed to realize the reflectivity measurements with smaller samples. The new slits are designed so that closest slit width in vertical direction is down to 10 μm, and the footprint on a sample will be reduced in 3 mm, about one tenth of that of ARISA-II, with a few percentages of q resolution. Here, a slit system with positioning sensors with high resistivity against radiation and repetitive reproducibility of less than 0.5μm (High-accuracy MT-touch Sensor, Metrol Co., Ltd.) was examined to obtain a slit width of 10 μm. Since typical sample size in the latest frontiers is around 10 mm × 10 mm, this is a great advantage to investigate new materials with small amount by a neutron reflectometer.

The replacement of old ARISA can realize reflectivity measurements in high-q region and with small samples. However, these measurements require long time due to low reflectivity and small beam size. To reduce the measurement time, we plan to introduce a focusing mirror (500 mm × 100 mm) with an ellipsoidal shape between the slits. By focusing the incident neutrons, on a sample, we can utilize high-flux neutron with large beam divergence, and it bring us the further reduction of measurement time and sample area. On the test measurements at ARISA-II, we succeeded in focusing neutron with 1 mm width at sample position, and the neutron flux at focusing position was about 10 times as much as a natural beam when aperture slit width was 0.4 mm [8]. Here we should keep it in mind that increase in beam flux is accompanied by worse angle resolution, because the beam flux is proportional to the product of its size and divergence according to Liouville's theorem. On the other hand, detector-focusing geometry allows us to obtain high intensity with keeping an angular resolution because the beam size at a sample can be enlarged. This is applicable for GISANS measurements by focusing in horizontal direction at the detector.

Furthermore, we are going to implement a double frame mode in the TOF system, that is, screening of the neutrons so that the frequency of neutron pulse (25 Hz) decreases into half (to 12.5 Hz) by adjusting the frequency of the disk chopper rotation after the neutrons coming in the beam line. The double frame mode is an attempt to expand the neutron's wavelength available for NR measurements into double, which the reflectivity with doubled q range can be obtained with an incident angle, although the duration time is also doubled. The double frame mode is of great use in the time-resolved NR measurements with wide q range. Our examination of the double frame mode on ARISA-II showed that it worked successfully and wavelength available was extended in double (from 2.0-8.8 Å to 2.0-17.6 Å). However, the disk chopper was made of aluminum plate coated by gadolinium paint which could not stop high energy neutrons generated at 25 Hz which cause a rise of background noise. To stop the fast neutrons, a T0 chopper made by Inconel block is going to be installed in back of the disk chopper.

5. Conclusions

In the present state of ARISA-II at BL-16 in MLF/J-PARC, it has achieved the off-specular reflection measurement and time resolved NR measurements as well as specular NR measurements from Soft-Interfaces. The time resolved NR measurements are brought by increment of the proton beam power, which provide us reduction of measurement time.

On the basis of the specification of ARISA-II, an advanced neutron reflectometer will be newly introduced at the beam line. The new reflectometer with enlarged vertical strokes of components is able to receive two downward beams (2.22 and 5.71 deg) for high q measurement (up to 0.56 Å^{-1}) on a liquid surface. The slit collimation is also improved to correspond to the smaller sample area down to 10 mm $\times$ 10 mm. Also, the T0 chopper and neutron focusing mirror are put into practical use in the new machine. The T0 chopper is able to stop the extremely fast neutrons which cause background noise. The sample focusing mirror brings us the further reduction of measurement time and sample area. The detector focusing mirror with a 2-dimensional detector can provide us high intensity with keeping an angular resolution in the GISANS measurement as an advanced measurement.

Acknowledgments

The neutron experiments were performed in S type research project, KEK (2009S08). The authors appreciate Mr. Setsuo Sato in KEK, Dr. Takeshi Fujiwara and Professor Hiroyuki Takahashi in Tokyo University for development of the 2-dimensional detector.

References

[1] Elman J F, Johs B D, Long T E and Koberstein J T 1994 *Macromolecules* **27** 5341.
[2] Bucknall D G, Higgins J S and Butler S A 2001 *Chem. Eng. Sci*, **56** 5473.
[3] Horinouchi A, Fujii Y, Yamada N L and Tanaka K 2010 *Chem. Lett.* **39** 810
[4] Takahara A, Terayama Y, Kikuchi M, Mitamura K, Kobayashi M and Yamada N L, J.Phys Conference Series, in press.
[5] Torikai N, Furusaka M, Matsuoka H, Matsushita Y, Shibayama M, Takahara A, Takeda M, Tasaki S and Yamaoka H 2002 *Appl. Phys., A* **74** S264.
[6] TerayamaY, Kikuchi M, Kobayashi M and Takahara A, Macromolecules, in press (DOI: 10.1021/ma102223y)
[7] Salditt T, Münster C, Mennicke U, Ollinger C and Fragneto G 2003 *Langmuir* **19** 7703.
[8] Torikai N, Yamada N L, Sagehashi H, Sugita T, Furusaka M, Higashi Y, Hino M, Fujiwara T and Takahashi H, *J. Phys.: Conf. Ser.* submitted.

Journal of Physics: Conference Series **272** (2011) 012018 doi:10.1088/1742-6596/272/1/012018

pH-Responsive Structural Change of PEGylated Amine-Bearing Nanogel Explored by Small Angle X-ray Scattering

Goshu Tamura [1,5], Yuya Shinohara [1,5], Isamu Akiba [2,5], Atsushi Tamura[3], Motoi Oishi[3], Yukio Nagasaki [3,4], Kazuo Sakurai [2,5] and Yoshiyuki Amemiya [1,5]

[1] Department of Advanced Materials Science, Graduate School of Frontier Sciences, The University of Tokyo, 5-1-5 Kashiwanoha, Kashiwa, Chiba 277-8561, Japan. [2]Department of Chemistry and Biochemistry, The University of Kitakyushu, Japan. [3]Department of Materials Science, Graduated School of Pure and Applied Science, University of Tsukuba, Japan. [4]Satellite Laboratory of International Centre for Materials Nanoarchitechtonics, National Institute for Materials Science, University of Tsukuba, Japan. [5]Japan Science and Technology Agency (JST) CREST

e-mail: goshu@x-ray.k.u-tokyo.ac.jp

Abstract. A PEGylated nanogel sphere containing cross-linked amino groups in its core undergoes dramatic volumetric changes upon pH-change. We measured small angle x-ray scattering from the nanogel and analyzed the data with a core-shell sphere model to determine the core size upon pH-change from 8.0 to 5.8. The protonation of the poly-amine core due to the pH-change increased the core volume by over 300%. Combining the obtained core volume and the absolute intensity, we estimated the change of the core electron density.

1. Introduction

Poly(ethylene glycol) (PEG) modification called "PEGylation" is covalent coupling of hydrophilic PEG to pharmaceutical materials to provide prolonged blood circulation time and biocompatibility[1]. The attached PEG forms an interface which can avoid unfavourable interactions between the coated material and other biological components such as enzymes. This technology was developed from pioneering work on chemical attachment of PEG to proteins [2], and then its efficacy has been proved on polymeric micelles constructed from PEG-hydrophobic block copolymers [3], where the hydrophobic chains aggregate to form a spherical core domain in aqueous solutions. Recently Nagasaki et al [4-8] developed a core-shell structured PEGylated nanogel consisting of a cross-linked poly[2-(*N,N*-diethylaminoethyl)methacrylate] (PEAMA) gel and tethered PEG chains. One advantage of this nanogel over polymeric micelles is that its core can swell drastically owing to the protonation of the amino groups. This volumetric change is so drastic as to be used for endosomal escape of the nanogel and pH-stimulated drug release.

Although the biological efficiency has been shown remarkably [7, 8], its physicochemical properties are not fully understood. Dynamic light scattering (DLS) studies have shown that the sphere size at lower pH is larger than that at higher pH [8]. This pH-dependent change is reversible and repeatable. Although it seems that the gel-core plays an important role in these responses, DLS doesn't provide detailed structural information about the core. Small angle x-ray scattering (SAXS) with synchrotron light sources is powerful tool to measure the core structure quantitatively. In the present study, we show the SAXS results on the structural change of the core of PEGylated nanogel with a pH-change.

2. Experimental

2.1. Materials

The PEGylated nanogel sphere, consisting of a PEAMA core cross-linked by ethylene glycol dimethacrylate (EGDMA) and a PEG shell, was synthesized with a previously reported method [8]. The weight-average molecular weight (M_w) of the PEG chain was 7870 g/mol and the polydispersity index (M_w/M_n) was 1.07, both being obtained by size exclusion chromatography [8], where M_n is the number-average molecular weight. The molar mixing ratio of EAMA, acetal-PEG, and EGDMA was 98.8: 1.2: 1.0. The solution pH was controlled at 5.8 and 8.0 by changing mixing ratio of 10 mM tris(hydroxymethyl)aminomethane (Tris) and 10 mM tris hydrochloride (Trizma) and was measured with a pH meter. Each solution of the sample was sonicated for 1 min using a homozinizer (SMT, UH-50) a few minutes before SAXS measurements. This treatment was found important to take out large aggregates that would cause to increase noises at low angles.

2.2. SAXS

SAXS experiments were carried out at BL03XU and BL45XU in SPring-8. An imaging plate [9] R-AXIS VII (RIGAKU, Japan) was used to measure the scattered x-ray intensity $I(q)$ as a function of the magnitude of the scattering vector q. Here, q is described by $q = (4\pi/\lambda) \sin \theta$, where 2θ is the scattering angle. Two ion chambers located upstream and downstream of the sample were used to measure x-ray transmittance of the sample. To reduce parasitic scatterings, we used a specially designed vacuum chamber [10]. The sample solutions (polymer concentration c = 2.0 mg/cm^3) were installed into a quartz capillary cell (φ = 2.0 mm, Hilegenberg GmbH, Germany) and then sealed with an epoxy resin. The sample to detector distances, wavelength of x-ray, and the exposure times were 4.0 and 3.5 m, 0.090 and 0.15 nm, and 30 and 300 s for BL03XU and 45XU, respectively. Wavelength and scattering vector were calibrated by the scattering of silver behenate. The obtained two-dimensional SAXS patterns of sample solutions were circularly averaged. Then experimentally determined scattering intensity profiles of the solvent were subtracted from obtained SAXS scattering profiles. Finally the differential scattering intensities were converted to the absolute intensity by use of water as a standard sample [10, 11]. SAXS intensity profiles at BL45XU (q-range: $0.05 < q < 0.5$ nm^{-1}) and BL03XU ($0.02 < q < 0.1$ nm^{-1}) were combined to cover a wider q-range.

SAXS experiments were performed at pH 5.8 and 8.0, since the previous DLS measurements [7, 8] showed a threshold of pH 7 in terms of hydrodynamic radius of the nanogel sphere.

To analyze the scattering data, we employed the theoretical scattering intensity of concentric core-shell spheres, which is given by:

$$I_{\mathrm{abs.}}(q) = Nr_e^2 \left\{\Delta\rho_C V_C F(q,R_C) + \Delta\rho_S V_S F(q,R_C+T_S)\right\}^2 \quad (1)$$

where N is the number of spheres per unit volume, and r_e is the classical electron radius, $\Delta\rho_C = \rho_C - \rho_S$ and $\Delta\rho_S = \rho_S - \rho_0$, where ρ_C, ρ_S, ρ_0 are electron densities of the core, the shell, and the solvent, respectively, Rc and Ts are the radius of the core and the thickness of the shell, and V_C and V_S are the volumes of the core and the shell, respectively. For spheres, the term of $F(q,R)$ is given by $3(\sin qR - qR\cos qR)/(qR)^3$. As the nature of the equation (1), we do not need the absolute intensity to determine Rc and Ts. For this case, N and $\Delta\rho$ are measured in arbitrary unit. Even when the absolute intensity is given, N and $\Delta\rho$ cannot be determined uniquely. In the fitting procedure we assumed that Rc had a Gaussian distribution in volume (δRc: variance of Rc). In addition, the effects of parasitic scatterings and dark noise of the detector were properly taken into account by adding an arbitrary constant value to equation (1).

3. Results

Figure 1 shows the measured SAXS intensity profiles and corresponding fitting curves at pH 5.8 and 8.0. It is clearly seen that there is a remarkable difference between the profiles at the different pH, showing that the size of the scattering object becomes larger with the decrease of pH. The fitting curve was obtained by the least square method using Igor Pro (WaveMetrics, USA). Due to the wide ynamic range (over three orders of magnitude) of $I(q)$, each data point was weighted by $I(q)^{-0.5}$. The fitting curves agree well with the experimental data, indicating that the present core-shell model is reasonable to describe the experimental result. The results of the fitting parameters are listed in table 1

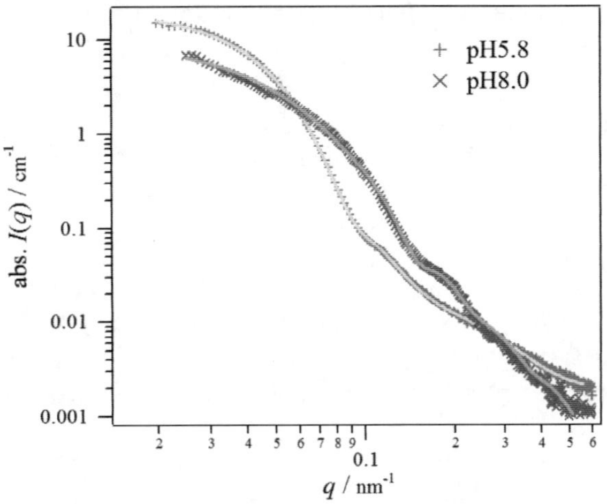

Figure 1. SAXS intensity profiles at pH 5.8 and 8.0, (red and blue marks, respectively), compared with the theoretical values calculated from the equation (1) with the parameter listed in table 1.

Table 1. Fitting parameters of the PEGylated nanogel sphere

	*R*c (nm)	*T*s (nm)	$\delta Rc/Rc$ (%)
pH 5.8	45	20	24
pH 8.0	30	42	20

As presented in table 1, *R*c increased by 50 % with the pH-change from 8.0 to 5.8, which means that the volume increased by over 300 %. By contrast, *T*s seemed to decrease by 50 % with the pH-change; this is reasonable since the core surface area increased by 2.3 times. It should be noted that the determination of *T*s may contain ambiguity owing to a small contrast between electron densities of water and PEG.

4. Discussion

It is important to determine the absolute values for ρ_C and ρ_S, since they can be related to the molecular composition. We know the value of ρ_0 for water: ρ_0 = 334 e/nm^3. After the intensities are converted into the absolute values, we still cannot determine ρ_C and ρ_S; we only have a multiplied value of $N(\Delta\rho_C)^2$. To resolve this problem, we used the following method; since V_C is given by SAXS, the total number of electrons in the core per unit volume is related by the following equation, if we can assume that the core at pH 8.0 does not include any water.

$$N\rho_C V_C = cP_C \frac{Ne_C N_A}{M_C} \tag{2}$$

where P_C is the weight ratio of the core to the whole sphere, N_A is Avogadro number, M_C is the molecular weight of the core, and Ne_C is the number of electrons in the core. P_C and the number of electrons per weight, $Ne_C N_A / M_C$, can be calculated from the composition ratio of the sample. For two unknown parameters, we have two sets of independent equations; therefore, we can uniquely determine N and ρ_C. In this way, we have determined N to be 1.0×10^{-8} [nm^{-3}] and the electron densities as shown in table 2. Here, the electron density of solvent was assumed to be the same with that of water, because the increase in electron density due to Cl^- in the buffer should be negligibly small.

The error of ρ_C can be evaluated from $\delta(\Delta\rho_C)/\Delta\rho_C$, which originates from $\delta Rc/Rc$. According to the error propagation law, $|\delta(\Delta\rho_C)/\Delta\rho_C|^2 = |\delta V_C/V_C|^2 = 3|\delta Rc/Rc|^2$, the errors of ρ_C are calculated as shown in table 2. We have evaluated the number of electrons transported into the core upon the pH-change on the basis of the changes of Rc and ρ_C. It is $(0.9\pm0.5) \times 10^8$ electrons. If we assume that these electrons are transported solely by water molecules, this value corresponds to $(0.9\pm0.5) \times 10^7$ of water molecules. It is roughly equal to the number of the water molecules in the core at pH 5.8, $(1.2\pm0.5) \times 10^7$, which is estimated by assuming that partial specific volume of water in the core at pH 5.8 is the same with that of water solution. This justifies our data analysis in the present study which is based on the assumption that the core at pH 8.0 does not include any water. It should be noted that the above values are obtained to only one significant figure, which is limited mainly by the dispersity of the nanogel spheres, $\delta Rc/Rc$. To be more precise in discussing the amount of the electron transport into/out of the core, it is required to prepare the sample which has a higher degree of monodispersity. Since $\Delta\rho_S$ is ~1/10 of $\Delta\rho_C$, the contribution of the shell to scattering intensity is so small that ρ_S is not evaluated preciously in the present study. The effort to enhance $\Delta\rho_S$ is under way.

To conclude, the present SAXS work is the first to quantitatively reveal the volumetric change of the core of the nanogel accompanied with pH-change and to propose a method to evaluate the absolute electron density of the core only based on the SAXS data and the core-shell model. The validity of the obtained electron density is planned to be examined combining other methods such as static light scattering.

Table 2. Evaluated electron densities of the core and the shell of the PEGylated nanogel sphere

	ρ_C (e/nm^3)	ρ_S (e/nm^3)
pH 5.8	344±3	335
pH 8.0	355±7	335

Acknowledgement
SAXS experiments were performed under the approval of the SPring-8 Proposal Advisory Committee (2009B1202 and 2009B7200). We appreciate SPring-8 beamline staff, Dr. Masunaga (BL03XU), and Dr. Ito (BL45XU) for their support in the experiments.

References
[1] Bontha S, Kabanov A V and Bronich T K 2006 *J. Control. Release* **114** 163.
[2] Monacelli R and Doretti M P 1964 *Boll. Lab. Chim. Prov. Bologna* **15** 235.
[3] Kataoka K, Kwon G S, Yokoyama M, Okano T and Sakurai Y 1993 *J. Control. Release* **24** 119.
[4] Oishi M and Nagasaki Y 2007 *React. Funct. Polym.* **67** 1311.
[5] Hayashi H, Iijima M, Kataoka K and Nagasaki Y 2004 *Macromolecules* **37** 5389.
[6] Oishi M, Hayashi H, Iijima M and Nagasaki Y 2007 *J. Mater. Chem.* **17** 3720.
[7] Oishi M, and Nagasaki Y 2010 *Nanomedicine* **5** 451.
[8] Tamura A, Oishi M and Nagasaki Y 2009 *Biomacromolecules* **10** 1818.
[9] Amemiya Y 1995 *J. Synchrotron Rad.* **2** 13.
[10] Akiba I, Terada N, Hashida S, Sakurai K, Sato T, Shiraishi K, Yokoyama M, Masunaga H, Ogawa H, Ito K and Yagi N 2010 *Langmuir* **26** 7544.
[11] Orthaber D, Bergmann A and Glatter O 2000 *J. Appl. Cryst.* **33** 218.

Journal of Physics: Conference Series **272** (2011) 012019 doi:10.1088/1742-6596/272/1/012019

Characterization of Swollen States of Polyelectrolyte Brushes in Salt Solution by Neutron Reflectivity

Motoyasu Kobayashi[1,2], Koji Mitamura[1,2], Masami Terada[2], Norifumi L. Yamada[3], Atsushi Takahara[*1,2,4]

[1] Institute for Materials Chemistry and Engineering, Kyushu University, 744 Motooka, Nishi-ku, Fukuoka 819-0395, Japan. [2] JST, ERATO, [3]High Energy Accelerator Research Organization (KEK), [4] Graduate School of Engineering, Kyushu University

takahara@cstf.kyushu-u.ac.jp

Abstract. Cationic and zwitterionic polyelectrolyte brushes on quartz substrate were synthesized by surface-initiated atom transfer radical polymerization of 2-(methacryloyloxy)-ethyltrimethylammonium chloride (MTAC) and 2-(methacryloyloxy)ethyl phosphorylcholine (MPC). The effects of ionic strength on brush structure are investigated by neutron reflectivity (NR) in NaCl deuterium oxide (D_2O) solutions. We observed that poly(MTAC) chains were drastically shrunk at concentrations above 0.1 M NaCl/D_2O, which may be the change in charge-screening effect against ions on poly(MTAC). On the other hand, effect of salt concentration on a swollen state of poly(MPC) brush was negligible, even at the high concentration (5.0 M) close to saturation. The behaviour of poly(MPC) in salt aqueous solution is completely different from that of poly(MTAC), which may arise from the unique interaction properties, neutral nature, and hydrated water structure of phosphorylcholine units.

1. Introduction

Polyelectrolyte brushes consist of polymer chains that are densely tethered one end to a surface while the other end is free, and their repeating units contain positive or negative charges [1]. In general, polyelectrolyte brushes have relatively stretched conformation in aqueous solution of low ionic strength, because of the intra- and intermolecular repulsive interactions and a higher osmotic pressure caused by the high ionic concentration inside the brush layer compared to outer bulk solution [2,3]. With increase in ionic strength, the electrostatic repulsion is screened by hydrated salt ions, and the difference in osmotic pressure of brush layer and bulk solution becomes smaller. As a result, polyelectrolyte brush behaves like an electrically neutral polymer forming a more coiled conformation, and the thickness of brush decreases. One of the typical examples is poly(2-(methacryloyloxy) ethyltrimethylammonium chloride) (poly(MTAC)) brush, whose conformational transition in aqueous NaCl solution was observed by Huck et al. using quartz crystal microbalance and scanning force microscopy [6]. On the contrary, phosphorylcholine-containing polyampholytes, such as poly(2-(methacryloyloxy)ethyl phosphorylcholine) (poly(MPC)), present an unusual phenomenon in salt solutions. Poly(MPC) is a super hydrophilic polyampholyte, which shows good blood compatibility, low friction, and anti-protein adsorption due to its cell membrane-like surface [7]. Recent dynamic light scattering study revealed that relatively expanded chain dimension of poly(MPC) in aqueous solution was hardly affected by salt concentration up to 1 M of NaCl [8], although the precise

Journal of Physics: Conference Series **272** (2011) 012019 doi:10.1088/1742-6596/272/1/012019

mechanism of this swollen behavior is still unknown. In this study, neutron reflectivity (NR) measurements were carried out to characterize the swollen structure of densely-grafted zwitterionic poly(MPC) brush on a flat substrate in aqueous solutions with much higher NaCl concentration than 1 M. The effect of salt concentration on the volume and thickness of cationic poly(MTAC) brush was also investigated as well as poly(MPC) brush by using 0 - 5 M of aqueous NaCl solution.

2. Experimental

Quartz plates (65 × 65 × 10 mm^3) were washed with piranha solution at 373 K for 1 h and subsequently exposed to vacuum ultraviolet-ray (λ = 172 nm) for 5 min under reduced pressure (30 Pa). The surface initiator, (2-bromo-2-methyl)propionyloxyhexyltrimethoxysilane, was immobilized on the substrate in toluene. Poly(MPC) brush was prepared by surface-initiated atom transfer radical polymerization in methanol using CuBr and 4,4'-dimethyl-2,2'-bipyridyl at 303 K for 12 h [9]. For MTAC polymerization, 2,2,2-trifluoroethanol was used as a solvent [10]. The number-average molecular weights (M_n) and molecular weight distribution of the free polymers simultaneously obtained with polymer brushes were determined by size-exclusion chromatography recorded on a JASCO instrument equipped with a JASCO 2031plus RI detector and TSK-gel column Super H6000, H4000, and H2500 using 0.01 M LiBr aqueous solution for poly(MPC) and 0.5 M of acetic acid aqueous solution containing 0.2 M sodium nitrate for poly(MTAC) as the eluents at a rate of 0.6 mL min^{-1} at 313 K. NR measurements were carried out by a time-of-flight (TOF) type reflectometer ARISA-II at BL16 in MLF/J-PARC providing 25 Hz-pulsed neutron radiation at 120 kW. Wavelength of the incident neutrons is tuned at 0.2 - 0.88 nm by a disk chopper. The neutron beam was irradiated through quartz at the interface between D_2O/ poly(MPC) brush on quartz glass and reflected neutrons were collected by a 2-D scintillation detector. The incident slits were adjusted to maintaining a 50 mm footprint size on the sample surface at all incident angles (typically 0.30, 0.75, 1.20 deg). The NR profiles were analyzed by fitting calculated reflectivity from the model scattering length density profiles to the data, using Parratt32 software.

3. Results and Discussion

We prepared poly(MTAC) with M_n = 94,300 in M_w/M_n = 1.17, and poly(MPC) with M_n = 137,000 in M_w/M_n = 1.85. The thickness under air (50% relative humidity at 25 °C) of the poly(MTAC) and poly(MPC) brushes on quartz substrates were 22.5 nm and 7.5 nm, respectively.

Figure 1 shows NR curves of poly(MTAC) brush interfaces in contact with D_2O solutions at several NaCl concentrations and the corresponding neutron scattering length density (SLD) profiles along with distance from the quartz surface. The volume fraction profiles in D_2O and 0.1 M D_2O solution against distance from the substrate are also represented in Figure 1 (g) and (h), respectively. In the case of D_2O/ poly(MTAC) interfaces, a critical angle was clearly observed at q = 0.11 nm^{-1} due to the difference in SLD between quartz and D_2O. The SLD profile of the poly(MTAC) brush in D_2O was a smooth upward curve from 4.6×10^{-4} to 6.3×10^{-4} nm^{-2} along with the distance from the substrate. The gradient profile indicated that the polymer chains in D_2O were stretched up to ca. 70 nm, which is thicker than 22.5 nm, the brush thickness in the air estimated by a NR measurement. Therefore, the polymer chains immersed in D_2O were extended significantly in the direction normal to the substrate. On the other hand, the volume fraction profile of poly(MTAC) in D_2O shows relatively high content of D_2O (volume fraction = 0.7) at the substrate, and polymer content decreased with distance from the substrate. It indicates that D_2O molecules deeply penetrated into the brush layer, resulting from the high osmotic pressure inside the brush. According to the self-consistent field theory, volume fraction (φ) of polymer brush with uniform chain length in a good solvent at a position (z) apart from the substrate is represented by parabolic function $\varphi(z) = \varphi_0(1-(z/h)^2)$ [11], where φ_0 is the volume fraction at z = 0, and h is the cut-off thickness of the swollen brush layer. In our case, the volume fraction profile in D_2O was in good agreement with a parabolic function with h = 75 nm. The estimation with the theoretical prediction indicates D_2O worked as a good solvent for poly(MTAC) brush, which resulted in the highly expanded brush structure.

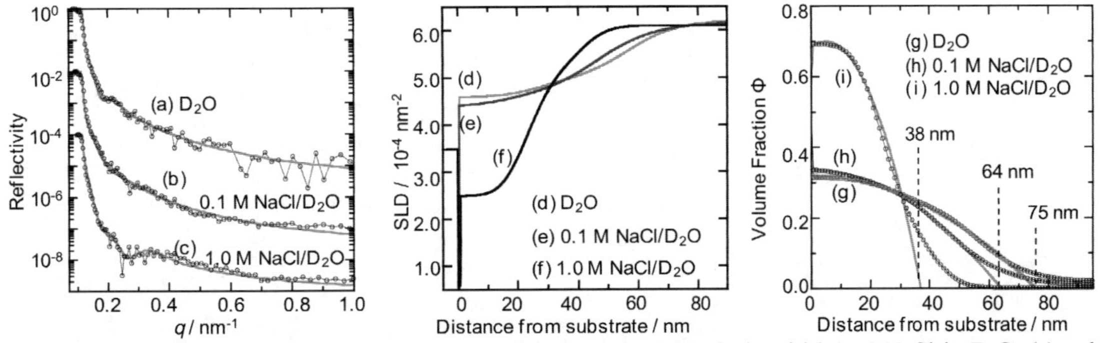

Figure 1. NR curves of (a) poly (MTAC) brush/D_2O, (b) poly (MTAC) brush/ 0.1 M NaCl in D_2O, (c) poly (MTAC) brush/1.0 M NaCl in D_2O, (d)-(f) their corresponding neutron SLD profiles along with the distance from quartz surface, respectively and (g), (h) volume fraction profiles of poly (MTAC) in D_2O and 0.1 M NaCl in D_2O with simulations using parabolic functions. Broken linear lines in (g)-(i) indicate swollen thickness h.

In the case of NaCl concentration of 0.1 M in D_2O, the SLD profile showed a similar feature to that in D_2O, that is, a gradual upward curve from 4.4×10^{-4} to 6.2×10^{-4} nm^{-2}. The volume fraction profile also suggests that a high content (65 % at the substrate) of D_2O exists in the brush layer. Additionally, the volume fraction is described well by the parabolic equation with h = 64 nm. These features are quite similar to that in D_2O, which indicates that the brush structure in 0.1 M NaCl/D_2O was still in a swollen state. In contrast, the SLD profile in [NaCl] = 1.0 M shows a significant reduction of SLD value at the substrate to 2.2×10^{-4} nm^{-2} and thickness of the brush layer to c.a. 45 nm, in comparison with those in D_2O and [NaCl] = 0.1 M. These facts suggest the shrinkage of the poly(MTAC) brush. We obtained a similar result in case of salt concentration of 5.0 M to that in [NaCl] = 1.0 M and confirmed that the brush had a shrunk conformation in such a concentrated condition.

Generally, the shrinkage or reduction of hydrodynamic radius of a free polyelectrolyte with positive or negative charges in an aqueous solution containing salt is thought as a transition from a swollen state to a more coiled state. The transition, which is resulted from screening of electrostatic repulsion between the monomer units by addition of salt, occurs at a critical point called "theta point", where the second virial coefficient is negligibly small. The brush's shrinkage observed in the poly(MTAC) brush system also took place by the same mechanism as shrunk behavior in the free polyelectrolyte system. Thus, a theta point should be also defined in the brush system and estimated between 0.1 -1.0 M in our experiment.

Figure 2 shows the NR curves of poly(MPC) in D_2O and 5.0 M NaCl/D_2O solution, and the corresponding SLD profiles. The volume fraction profile in D_2O was also shown in Figure 2(e). The SLD profile of the brush in D_2O monotonically increases with the distance from the substrate and intersects with the SLD of the solvent at c.a. 60 nm. Taking dry thickness (7.5 nm at 50 % relative humidity at 25°C) into account, it indicates that the poly(MPC) brush has a stretched conformation in a D_2O medium. However, in contrast to poly(MTAC) brush, the volume fraction profile of poly(MPC) brush in D_2O did not describe by simple parabolic function representing swollen brush structure in a good solvent. This discrepancy could be explained by molecular weight distribution of the polymer brush since the parabolic expression does not include the effect of heterogeneity of chain length. Indeed, poly(MPC) has a larger polydispersity index (M_w/M_n = 1.85) than that of poly(MTAC) (M_w/M_n = 1.17), the disagreement with the theoretical prediction was dominant in the case of poly(MPC). The monotonically decreasing volume fraction as shown in Figure 2(e) does not conflict with another theoretical prediction considering the molecular weight distribution effect reported by other groups, where an exponential-like decay appears in a volume fraction profile of polymer brush with large molecular weight distribution in a good solvent [12].

On the other hand, in the case of poly(MPC) brush in [NaCl] = 5.0 M (near saturation at 25 °C), the SLD profile was almost the same as that in D_2O (see Figure 2(c) and (d)). A similar result was also obtained in more concentrated condition at 1.0 M. These facts suggest the swollen states of poly(MPC

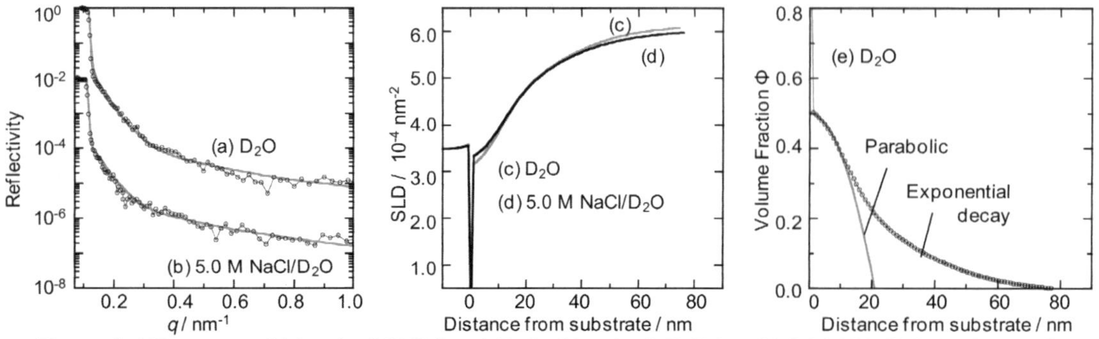

Figure 2. NR curves of (a) poly (MPC) brush/D_2O, (b) poly (MPC) brush/ 5.0 M NaCl in D_2O, (c)-(d) their corresponding neutron SLD profiles along with the distance from quartz surface, respectively, and (e) volume fraction profile of poly (MPC) brush in D_2O with a simulation using parabolic function.

brush is independent of salt concentration, in contrast to the poly(MTAC) brush, which indicates that the poly(MPC) should have no transition, i.e., it has no theta salt concentration point. The origin of this characteristic behavior of poly(MPC) brush in aqueous solutions of NaCl is not clear but might come from its unique interaction among the brush side chains, its neutral nature or structure of hydration around the brush.

4. Conclusion

In this work, we have investigated on swelling behaviors of poly(MTAC) and poly(MPC) brushes in NaCl D_2O solutions using neutron reflectometry. It was revealed that the brush of poly(MTAC) shrunk drastically at NaCl concentration between 0.1 and 1.0 M, which corresponds to theta point. We also found that the brush structure of poly(MPC) is not affected by the NaCl aqueous solutions contacting with the brush. The unique behavior of poly(MPC) brush in a D_2O solution of NaCl might be closely related with its properties in tribology, anti-fouling, and wettability.

Acknowledgments

The present work was supported in the S-type research project in KEK (2009S08).

References

[1] Rühe J, Ballauff M, Biesalski M, Dziezok P. Gröhn F, Johannsmann D, Houbenov N, Hugenberg N, Konradi R, Minko S, Motornov M, Netz R R, Schmidt M, Seidel C, Stamm M, Stephan T, Usov D and Zhang H 2004 *Adv. Polym. Sci.* **165** 79.
[2] Zhulina E B, Wolterink J K and Borisov O V 2000 *Macromolecules* **33** 4945.
[3] Luzinov I, Minko S and Tsukruk V V 2004 *Prog. Polym. Sci.* **29** 635.
[4] Takahashi A, Kato N and Nagasawa M 1970 *J. Phys. Chem.* **74** 944.
[5] Nagasawa M and Eguchi Y 1976 *J. Phys. Chem.* **71** 880.
[6] Moya S E, Azzaroni O, Kelby T, Donath E and Huck W T S 2007 *J. Phys. Chem. B* **111** 7034.
[7] Ishihara K, Ueda T and Nakabayashi N 1990 *Polym. J.* **22** 355.
[8] Matsuda Y, Kobayashi M, Annaka M, Ishihara K and Takahara A 2008 *Langmuir* **24** 8772.
[9] Kobayashi M, Terayama Y, Hosaka N, Kaido M, Suzuki A, Yamada N, Torikai N, Ishihara K and Takahara A 2007 *Soft Matter* **3** 740.
[10] Kobayashi M, Terada M, Terayama Y, Kikuchi M and Takahara A 2010 *Macromolecules* **43** 8409.
[11] Milner S T, Witten T A and Cates M E 1988 *Macromolecules* **21** 2610
[12] de Vos W M, Leermakers F A M, de Keizer A, Kleijn J M and Cohen Stuart M A 2009 *Macromolecules* **42** 5881.

Journal of Physics: Conference Series **272** (2011) 012020 doi:10.1088/1742-6596/272/1/012020

X-ray Photon Correlation Spectroscopy of Silica Particles Grafted with Polymer Brush in Polystyrene Matrix

Taiki Hoshino[1,2], Moriya Kikuchi[1,2], Daiki Murakami[1,2], Koji Mitamura[1,2], Yoshiko Harada[1,2], Kiminori Ito[2], Yoshihito Tanaka[2], Sono Sasaki[2,3], Masaki Takata[2], Atsushi Takahara[1,2,4]

[1]JST,ERATO, 744 Motooka, Nishi-ku, Fukuoka 819-0395, Japan. [2]RIKEN SPring-8 Center, Harima Institute, [3]Graduate School of Science and Technology, Kyoto Institute of Technology, [4]Institute for Materials Chemistry and Engineering, Kyushu University

E-mail: t-hoshino@cstf.kyushu-u.ac.jp

Abstract. X-ray photon correlation spectroscopy system was setup at SPring-8, BL19LXU, and the partial coherence scattering data from the silica particles grafted with polymer brush in polystyrene matrix were measured. Firstly, the static speckle patterns were checked. Below the glass transition temperature of polystyrene (T_g), speckles were clearly observed, on the other hand, above T_g, the scattering patterns became smooth and speckles were hardly observed. These variances of the speckle patterns result from the particle motion. Secondly, from the time variance of the speckle data, time autocorrelation functions $g_2(q,t)$ are calculated. While the flat behaviour of $g_2(q,t)$ without relaxation were observed below T_g, the relaxation behavior with relaxation time~10^0-10^1 were observed above T_g

1. Introduction

Static and dynamic structures of nanoparticles in soft materials have received lots of interests from various view points. Studies using x-ray, small angle x-ray scattering for statics and x-ray photon correlation spectroscopy (XPCS) for dynamics [1,2,3,4], have been performed for investigating various interesting properties such as reinforcement of polymers or hyper diffusive motion near glass transition.

XPCS has been regarded as a powerful tool for revealing the dynamic structure of soft materials and the measurements are now routinely performed at ID1-10A of ESRF (France) and 8-ID of APS (U.S.A) using third-generation synchrotrons. But there is no dedicated beamline for XPCS in SPring-8, and very few papers [5] are found for performing XPCS. In this work, XPCS system was set up at SPring-8, BL19LXU (Hyogo, Japan), and the dynamic behavior of the polystyrene (PS) -grafted silica particles in PS matrix was investigated.

2. Experimental

The measurement was carried out at 27-m-long undulator bemline BL19LXU in SPring-8. Details of the beamline optics are presented elsewhere [6]. Figure 1 shows a schematic layout of the setup. The undulator source and the monochromator were tuned to an energy of 7.25 keV ($\lambda = 1.71$ Å) and higher harmonic x-rays were removed by the Pt-coated mirrors. The energy bandwidth of the Si(111)

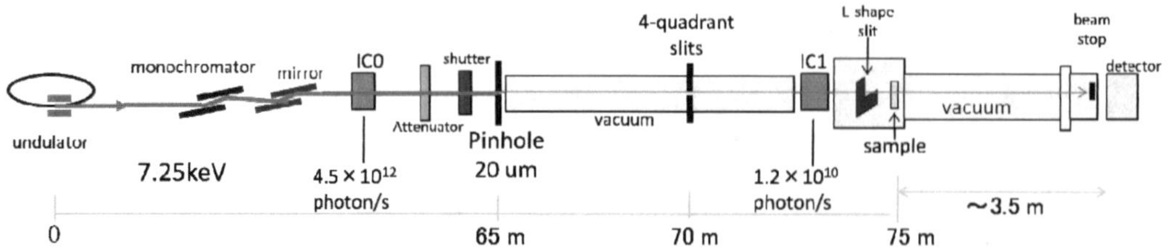

Figure 1. Schematic drawing of the experimental setup for XPCS at SPring-8, BL19LXU.

monochromator leads to a longitudinal coherence length $\xi_l = \lambda^2 / \Delta\lambda \sim 1\,\mu\text{m}$. The beam size of the light source was $\sigma \sim 113 \times 14 (\text{H} \times \text{V})\,\mu\text{m}^2$, and the distance between the light source and the sample was *R*~75 m. By calculating from the equation $\xi_t = \lambda R / (2\pi\sigma)$, the transverse coherence length at the sample was to be obtained $\xi_t \sim 18 \times 146 (\text{H} \times \text{V})\,\mu\text{m}^2$. To ensure enough transverse coherence, we set a 20 μm diameter pinhole upstream 10 m of the sample. The parasitic scattering was cut by the 4-quadrant slits and the L-shaped slit, and the partially coherent x-ray was illuminated to the sample in vacuum. The typical flux at the sample was $\sim 1 \times 10^{10}$ photons/sec. The scattered x-rays were detected ~3.5 m downstream using a direct illuminated charge-coupled device (CCD) camera (PI-LCX 1300, Princeton Instruments) with 1340×1300 pixels and a pixel size of $20 \times 20\,\mu\text{m}^2$. In this setting, a scattering vector of $2.4 \times 10^{-2} < q < 4 \times 10^{-1}\ \text{nm}^{-1}$ was covered, where $q = (4\pi / \lambda)\sin\theta$ is the magnitude of the scattering vector, where λ and 2θ are the wavelength of X-rays and scattering angle, respectively. This *q*-range was small enough compared to the maximum of $q \sim 1.6\ \text{nm}^{-1}$ estimated from the requirements for XPCS that ξ_l should be smaller than the path difference. The time variance of the speckle data are measured in *kinetics mode* typically using 10 pixels wide window with ~130 frames per 300 and 1000 ms.

The sample preparation of PS-grafted silica particles in PS matrix will be briefly presented here [7]. The surface-initiator (2-bromo-2-methyl) propionyloxyhexyltriethoxysilane (BHE) was immobilized on silica nanoparticle (108 nm diameter; Nissan Chemicals) by solution method. Subsequently, polystyrene was grafted on silica nanoparticle using surface-initiated atom transfer radical polymerization with the ligand of N,N,N',N",N"-pentamethyldiethylenetriamine in the solution of dimethylacetamide. The molecular weight of the grafted PS was $M_n = 1.23 \times 10^5$ g/mol, M_w/M_n=1.67, and the graft density was 0.95 chains/nm^2. These particles were dispersed in polystyrene matrix

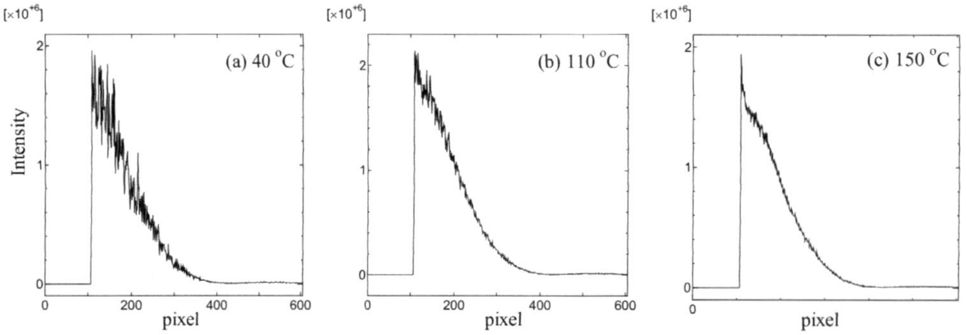

Figure 2. Scattering intensity profiles, obtained by slicing the 2-dimensional CCD images with total measurement time of ~2 minutes at (a) 40 °C, (b)110 °C and (c) 150 °C.

($M_w = 5.40\times10^4$, M_w/M_n=1.04). The content of the pure silica was 0.5 wt%. The possibility of aggregation of nanoparticles was checked by small angle scattering down to $q = 2.4\times10^{-2}$ nm^{-1}, near Guinier region, any aggregation features, such as intensity increasing at low-q region, are not observed.

3. Results and Discussion

Firstly the temperature dependence of speckle pattern is presented here. Figure 2 shows the scattering intensity profiles, obtained by slicing the 2-dimensional CCD images accumulated 120 times per exposure time 500 ms (total measurement time ~2 minutes) at 40 °C (a), 110 °C (b) and 150 °C (c). The profile at 40 °C (a) shows jaggy "speckle" pattern, which results from that the coherent scattered x-rays interfered at the detector. But the profiles at 110 °C (b) and 150 °C (c) are smooth and speckles are hard to be observed. These variances of the speckle patterns with increasing temperature result from the thermal motion of the silica particles as well as micro-Brownian motion of matrix PS. The silica particles in polystyrene were immobile below the glass transition temperature (T_g) of polystyrene (~100 °C) and the speckles from the silica particles were clearly observed. But, above T_g, the speckles moved because of the thermal motion of the silica particles, and the speckles were averaged during the measurement time.

The time-autocorrelation function $g_2(q,t)$ for each pixels are calculated from the speckle data by

$$g_2(q,t) = \langle I(q,t')I(q,t'+t)\rangle / \langle I(q,t')\rangle^2, \quad (1)$$

where $I(q,t')$ is the scattered intensity at wave vector q and time t'. The calculated $g_2(q,t)$ are averaged over finite-sized regions in reciprocal space. The data at q=3.5×10^{-2} nm^{-1}, at 40 °C (a), 110 °C (b) and 150 °C (c) are shown in Fig. 3 as representative data for the temperature dependence. At 40 °C, $g_2(q,t)$ with sampling time of 300 ms and 1000 ms are flat around the value of unity. Same behavior is also shown at every q at 40 °C. These results indicate that the sample is in static state at 40 °C (<T_g) and the period over which the synchrotron and our setup are sufficiently stable. On the other hand, at 110 and 150 °C, $g_2(q,t)$ shows non-flat behavior. Since these temperatures are higher than T_g, these behavior are considered to reflect the motion of the silica particles. Autocorrelation functions for soft glassy material are often by the stretched exponential expression,

$$g_2(q,t) = A\exp[-2(t/\tau)^\beta]+1, \quad (2)$$

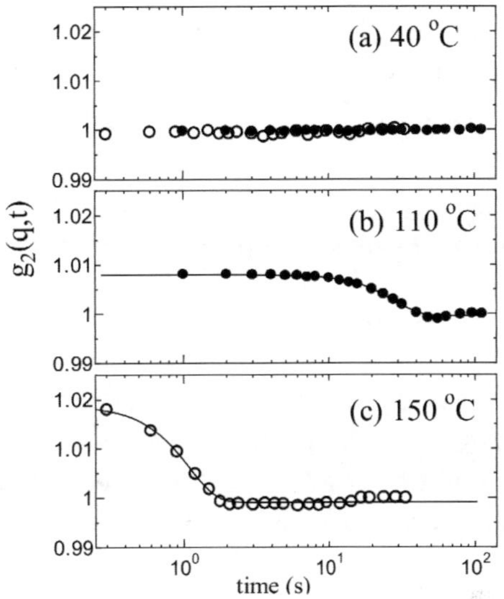

Figure 3. Autocorrelation functions $g_2(q,t)$ at q=3.5×10^{-2} nm^{-1}. The measured temperatures are (a) 40 °C, (b) 110 °C and (c)150 °C. Open circles and solid circles are corresponding to the sampling time of 300 and 1000 ms respectively. Solid lines in (b) and (c) are fitting curves by Eq. (2).

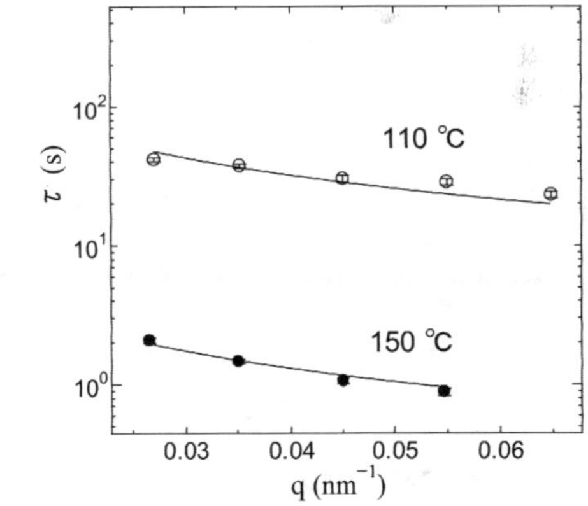

Figure 4. Relaxation time τ plotted against q for 110 °C (open circles) and 150 °C (solid circles). The solid curves are fitting curves by $\tau = A/q$. The dashed line calculated from simple Brownian behaviour model.

Journal of Physics: Conference Series **272** (2011) 012020 doi:10.1088/1742-6596/272/1/012020

where β quantifies the deviation from exponential relaxation, $\beta = 1$ corresponding to the behavior of particles like noniteracting Brownian motion in a Newtonian liquid, and τ the relaxation time. The prefactor A, speckle contrast, is small compared with the ideal value of 2. This speckle contrast depends on the values of the beam's coherence, the scattering vector and so forth. Calculated from the static speckle patterns, the speckle contrast at $q = 3.5\times10^{-2}$ nm^{-1} is ~1.03. This is fairly consistent with the value of A. But the small value compared with 1.03 may indicate the existence of faster mode. By fitting measured $g_2(q,t)$ at 3.5×10^{-2} nm^{-1} by Eq. (2), $\beta = 2.44, \tau = 37.5$s at 110 °C and $\beta = 2.26$, $\tau = 1.47$s at 150 °C are obtained. At other q ($2.6\times10^{-2} < q < 6.5\times10^{-2}$ nm^{-1}), $\beta > 1$ are also obtained. The relationship $\beta > 1$ in equilibrium system in "soft solids" has been indicated by many experiments [1,2,3,4], and such behavior is considered to be resulted from hyperdiffusion or ballistic motion. In the literatures, the relationship $\beta < 2$ is common feature. Compared with the values in the literatures, the present values are a little larger. The present observation may indicate the effect of faster modes or another kind of motions, now the possibilities of the sedimentation or the flowing phenomena of the particles are not ruled out. In Fig. 4, τ against q at 110 and 150 °C are plotted. There is a slight decrease with increasing q both at 110 and 150 °C. The fitting procedure by $\tau = C/q$, which is sometimes performed for the data with $\beta > 1$, is done as shown in Fig.4. The fitting results look sufficient, but the validity of the fitting is not argued here because the present data may not include enough number of plots.

4. Conclusion

XPCS system was set up at SPring-8, BL19LXU, and the dynamics of the polystyrene grafted silica particles in polystyrene matrix was measured. Firstly, the static speckle patterns were checked with total measurement time of ~2 minutes. At 40 °C (<T_g), speckles were clearly observed, but, at 110 and 150 °C (>T_g), the speckles were disappeared and the smooth intensity patterns were observed. These variations of the speckle patterns result from the particle motion. Secondly, from the time dependence of the speckle data, time autocorrelation functions $g_2(q,t)$ are calculated. In $g_2(q,t)$, while flat behavior without relaxation were observed below T_g, the relaxation behavior with $\tau \sim 10^0$-10^1 were observed above T_g. In the present work, we successfully observed the dynamics of the silica particles in PS matrix by XPCS, but the more measurement data are necessary for the detailed discussions.

The synchrotron radiation experiments were performed under the approval of RIKEN (Proposal No. 20100039).

References

[1] Narayanan R A, Thiyagarajan P, Lewis S, Bansal A, Schadler L S, and Lurio L B 2006 *Phys. Rev. Lett.* **97** 075505.
[2] Caronna C, Chushkin Y, Madsen A and Cupane A 2008 *Phys. Rev. Lett.* **100** 055702.
[3] Guo H, Bourret G, Corbierre M K, Rucareanu S, Lennox R B, Laaziri K, Piche L, Sutton M, Harden J L, and Leheny R L 2009 *Phys. Rev. Lett.* **102** 075702.
[4] Akcora P and Kumar S K; Moll J; Lewis S and Schadler L S; Li Y and Benicewicz B C; Sandy A, Narayanan S, Ilavsky J and Thiyagarajan P; Colby R H; Douglas J F 2010 *MacroMolcules* **43** 1003
[5] Shinohara Y, Kishimoto H, Maejima T, Nishikawa H, Yagi N, and Amemiya Y 2007 *Jpn. J. Appl. Phys.* **46** L300.
[6] Yabashi M, Mochizuki T, Yamazaki H, Goto S, Ohashi H, Takeshita K, Ohata T, Matsushita T, Tamasaku K, Tanaka Y, Ishikawa T 2001 *Nucl. Instr. and Meth. A* **467-468** 678.
[7] Matsuda Y, Kobayashi M, Annaka M, Ishihara K, and Takahara A 2008 Langmuir 24, 8772.

Journal of Physics: Conference Series **272** (2011) 012021 doi:10.1088/1742-6596/272/1/012021

Molecular Aggregation States of Imogolite/P3HT Nanofiber Hybrid

Weng On Yah,[1] Atsushi Irie,[1] Hideyuki Otsuka,[1,2] Sono Sasaki,[3] Naoto Yagi,[3] Masugu Sato,[3] Tomoyuki Koganezawa,[3] Atsushi Takahara[1,2,*]
[1]Graduate School of Engineering, [2]IMCE, Kyushu University, [3]Japan Synchrotron Radiation Research Institute.

E-mail: takahara@cstf.kyushu-u.ac.jp

Abstract. Preparation of novel organic/inorganic nanofiber hybrid composed of one-dimensional P3HT nanofiber and aluminosilicate imogolite was demonstrated. To achieve better compatibility with hydrophobic P3HT, Al-OH group on imogolite surface was modified with terthiophene derivative, 2-[5"-hexyl-2,2':5',2"-terthiophene]-5-ethyl)-phosphonic acid (H3TP). Dissolving the mixture of P3HT and surface modified imogolite in poor solvent of anisole at elevated temperature (> 70°C) followed by subsequent slow cooling (25°C/h) to room temperature induced the formation of P3HT crystallite that winds around the surface modified imogolite core. The thickness of the nanofiber hybrid estimated from DFM is larger than the parent P3HT nanofiber and imogolite. UV-vis spectroscopy spectra indicate that addition of surface modified imogolite can influence the molecular aggregation state of P3HT via van der Waals interaction. Through GIWAXD experiments, it was found that P3HT molecular chains are strongly oriented and lying on surface modified imogolite core.

Introduction

Good solvent processibility of hexyl side chain poly(hexylthiophene) (P3HT) which allowed for fabrication of nanofibers [1] and thin films has attracted much research interest both in academia and industry area. Intrinsic nanofiber formation due to the introduction of chain alignment and extension will lead to significant improvements in electrical and mechanical properties [2,3]. Various physicochemical aspects such as synthesis [4], electrical [3], mechanical [5], thermochromism, solvatochromism [6], and crystal structure [7] have been intensively studied. Nevertheless, relatively little attention has been paid to the molecular aggregation structures of poly(3-alkylthiophene)s nanofibers, particularly when it comes to hybridization with inorganic nanofiller. In modern plastics technology, polymer reinforced with fillers is considered as a common way to achieve enhanced properties. The nanofiller have at least one characteristic length scale of nanometer order, which is about the size of the radius of gyration of a polymer chain. Thus, it is expected that the nanofiller can alter the physicochemical properties of the polymeric nanofiber much more significantly than in the bulk polymer matrix. Clay and carbon nanotubes [8] are widely used as nanoscale filler in fabricating polymer nanocomposites. However, carbon nanotube and some inorganic nanotubes are artificially materials involving expensive and time-consuming preparation processes. Additionally, these materials are opaque due to conjugated π-system of carbon nanotube and energy band gap of some transition metals which is not suitable for certain optical application. Here, high aspect ratio of the imogolite, a hydrous aluminosilicate ($Al_2O_3•SiO_2•H_2O$) hollow nanotube material with its unique

Journal of Physics: Conference Series **272** (2011) 012021 doi:10.1088/1742-6596/272/1/012021

nanostructure was proposed as the nanofiller in P3HT nanofiber for electronic and optical applications. Imogolite have shown great potential as one-dimensional nanofiller due to their small diameter, good mechanical properties and potential for chemical functionalization. Reinforcement of P3HT nanofiber by imogolite is expected to improve mechanical and thermal stability as well making the resulted hybrid material endure to outer environment more persistently. Since the imogolite surface consisted of hydrophilic Al-OH groups, there is a necessity to increase the imogolite hydrophobicity via surface modification with alkyl phosphonic acid group for improved compatibility with P3HT nanofiber. To explore their reinforcing potential, molecular aggregation states and molecular orientation of P3HT chain on imogolite were investigated.

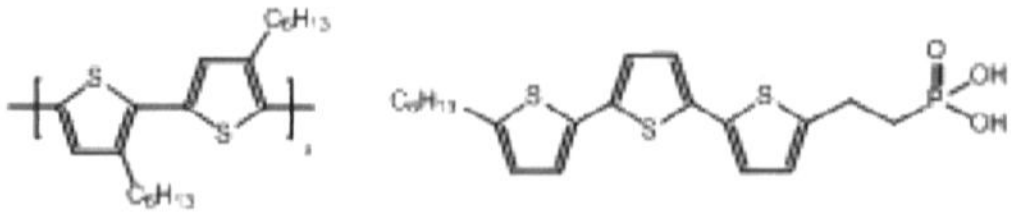

Figure 1. Chemical structure of regioregular P3HT (left) and HT3P (right)

1. Experimental

For preparation of a P3HT nanofiber solution, a P3HT with chemical structure as shown in Figure 1 (1.0 mg) was added to anisole (2.0 g) in a 5 mL glass vial, and the solution was stirred with heating (> 70 °C). After the stirring was stopped at 70 °C, the P3HT solutions was gradually cooled to 20 °C with cooling rate of 25 °C/h . Nanofibers mainly grow in the solution during cooling [9]. Imogolite prepared according to the method reported by Farmer *et al.* [10] was surface modified with terthiophene derivative, 2-[5"-hexyl-2,2':5',2"-terthiophene]-5-ethyl)-phosphonic acid (H3TP) with chemical structure as shown in Figure 1. Typically, H3TP modified imogolite was obtanied by mixing the imogolite with ethanol solution of H3TP, centrifugation, rinsing in ethanol and freeze dried. H3TP modified imogolite was added to the P3HT solution in anisole (conc. 0.0005 wt%, H3TP modified imogolite:P3HT = 1:1), heated to 70°C overnight, cooled to room temperature (25°C/h) to produce H3TP modified imogolite/P3HT nanofiber hybrid (Figure 2). The nanofiber solution was spin coated at 3000 rpm on SiO_2/Si substrate. The nanofibers deposited on the substrate were imaged in air using DFM SII SPA 400. UV-vis absorption spectra were recorded on a Shimadzu UV-3600 spectrophotometer. GIWAXD measurements were carried out at a BL46-XU beamline of Spring-8 (Japan Synchrotron Radiation Research Institute, Hyogo, Japan). The wavelength, λ, of monochromatized incident X-rays was 0.1 nm. The data collection time was 3.0 s per step, and the angular interval between steps was 0.05°. Figure 3 shows the schematic geometry of the in-plane and out of plane GIWAXD measurement. When the incident angle of X-ray, α_i, is equal to, or smaller than critical angle, α_c, the incident angle X-rays undergo total external reflection and penetrate into the sample as evanescent waves. The Bragg diffractions from crystallographic planes perpendicular and parallel to the film surface were obtained from surface and bulk regions at the incident angle of X-rays, α_i, of 0.08° and 0.16°, respectively [11].

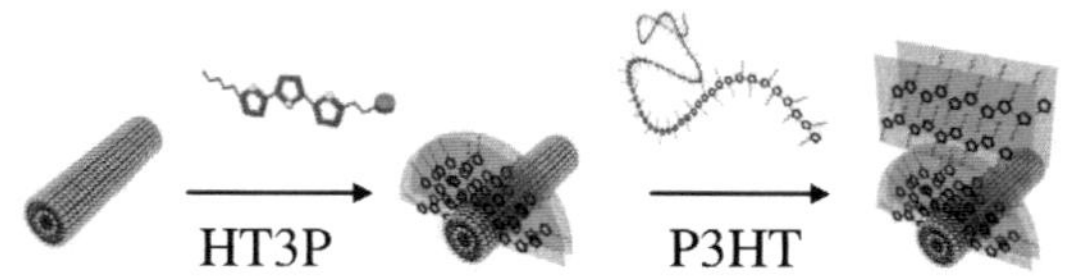

Figure 2. Fabrication and postulated molecular model of imogolite/P3HT nanofiber hybrid.

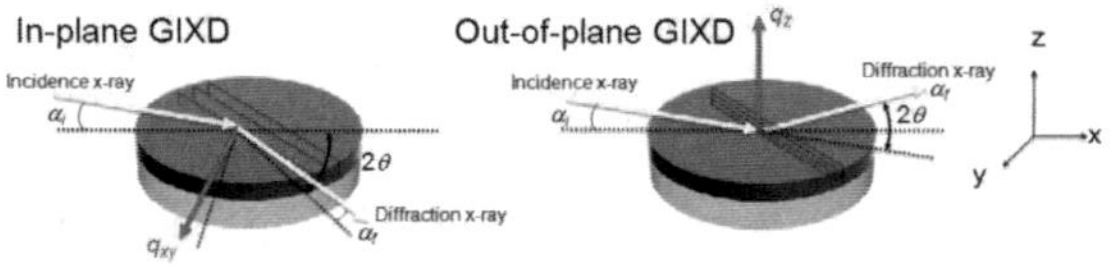

Figure 3. Schematic geometry of the in-plane (left) and out of plane (right) GIWAXD measurements.

2. Results and Discussion

Generally, P3HT can be dissolved well in good solvent such as chloroform or toluene at room temperature. Rapid evaporation of the solvent will result in solidification of P3HT from solution forming weakly crystalline solid without perceptible morphological structure [12, 13]. Unlike the above commonly used experimental conditions, P3HT was found to form well defined fibrous like crystalline structure when P3HT dissolved in elevated temperature (> 50 °C) of poor solvent such as anisole or xylene was cooled to room temperature. Representative example of P3HT nanofiber is shown in Figure 4 (b). Colour changed from the original orange solutions to dark red which referred to as thermochromism occurred during the cooling process [14]. Typically, the width and length of the P3HT nanofiber were found to be of the order of ca. 15 nm and 1 μm, respectively. The thickness, as estimated from height image of DFM, was in the order of ca. 5 nm. Imogolite as shown in Figure 4 (a) exhibited rigid fiber like morphology. The average length of the synthesized imogolite was several hundred nanometers. The heights of nanofiber hybrid observed were 2 to 3 times larger than the diameter of single imogolite nanotube indicating the formation of bundle of imogolite nanofiber [15]. On the other hand, the nanofiber hybrid as shown in Figure 4 (c) exhibited a swollen fibrous morphology compared to their parent imogolite and P3HT. This swollen morphology was due to the intertwining effect of combining rigid like imogolite bundle with entangled web like structure of P3HT nanofiber. Figure 5, shows the temperature dependent UV-vis absorption spectra of P3HT solution and imogolite/P3HT nanofiber hybrid. At higher temperature, both spectra exhibited a single peak at λ_{max} of 450 nm similar to P3HT solution in good solvent of chloroform indicating their complete dissolution in anisole. During the cooling to room temperature, intensity of λ_{max} at 450 nm was decreased in both spectra with appearance of vibronic structure at longer wavelengths (500-650 nm). Isobectic point were observed around 489 nm and 525 nm in the absorption spectra of P3HT solution and imogolite/P3HT, respectively, recorded in a temperature range of 20-70 °C. The isobectic point observed suggests that both P3HT and imogolite/P3HT exhibited two different states in a solution, namely isolated coil-like conformation polymer chains which dissolved completely in a solution and the other state is rod like conformation of polymer aggregate dispersed in the solution, which corresponds to the one-dimensional nanofibrillar structure as visualized by DFM [9]. Shift of isobectic point and lower coil like conformation absorption intensity upon hybridization with imogolite showed

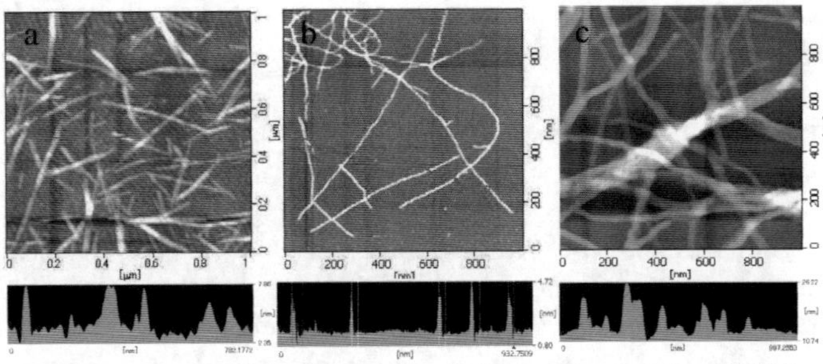

Figure 4. DFM images of imogolite (a), P3HT nanofiber (b) and imogolite/P3HT nanofiber hybrid (c).

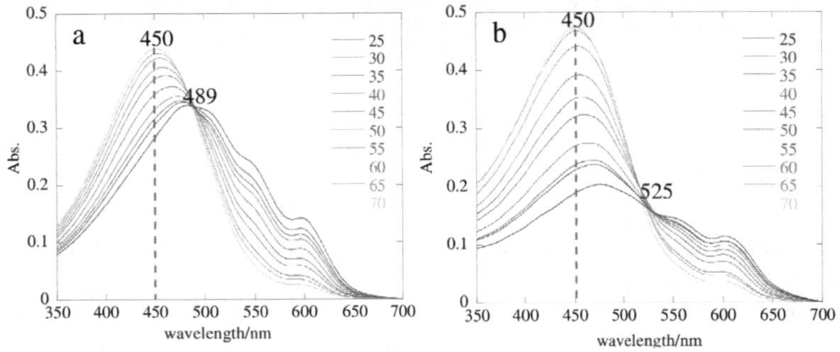

Figure 5. Absorption spectra of P3HT nanofiber (a) and imogolite/P3HT nanofiber (b).

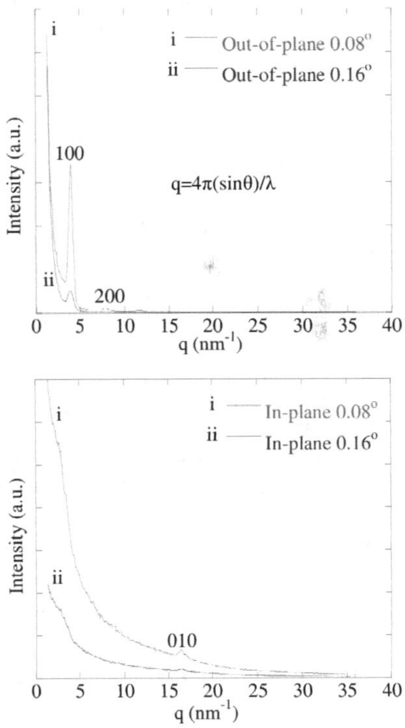

Figure 6. In-plane and our-of-plane GIWAXD patterns of imogolite/P3HT nanofiber hybrid.

the presence of interaction between imogolite with P3HT polymer chain. As suggested from the DFM and UV-vis results, P3HT chains co-exit with rod like imogolite, winding around it by van der Waals interactions causing in swelling structure of the imogolite/P3HT nanofiber hybrid compared to the pristine P3HT nanofiber and imogolite. The orientation of P3HT crystalline phase on imogolite surface via van der Waals interaction was investigated by GIWAXD method [16,17] as shown in Figure 6. In the out-of-plane GIWAXD pattern, the peaks at d = 4 (100) and 7.8 nm (200) were corresponding to the ordering of the P3HT hexyl side chain. P3HT nanofiber showed the intensity which suggested higher crystallinity on the surface ($\alpha_i = 0.08°$) compared to the deeper region ($\alpha_i = 0.16°$) where the semi crystalline imogolite mostly reside. Similar trend was observed also in the in-plane GIWAXD pattern with π-π* stacking of P3HT thiophene ring. This trend is much obvious at the surface region ($\alpha_i = 0.08°$) compared to the bulk region ($\alpha_i = 0.16°$). These results indicate that P3HT chains are winding around imogolite nanotube along the imogolite axis in which the P3HT acting as sheath lie on the top of imogolite core once the nanofiber hybrid deposited on the substrate.

3. Conclusion

In this study we have shown that the hydrid nanofiber of imogolite/P3HT can be prepared by readily precipitation from relatively poor solvent of anisole. Surface modified imogolite had a great influence on the molecular structure of P3HT nanofiber. The thickness of imogolite /P3HT nanofiber hybrid as estimated from the height image of DFM is much larger than that of parent P3HT nanofiber. UV-vis spectra showed interaction presence between the P3HT molecular chain and surface modified imogolite. GIWAXD studies revealed that P3HT molecular chain are strongly oriented, namely, “lying” above the imogolite core on the substrate.

Acknowledgements
This work was supported by a Grant-in-Aid for the Global COE Program, “Science for Future Molecular Systems” (No. 1520258), and Grant-in-Aid for Scientific Research A (No. 19205031) from the Ministry of Education, Culture, Science, Sports and Technology of Japan.

References

[1] Ihn K J, Moulton J and Smith P 1993 *J. Polym. Sci., Part B: Polym. Phys.* **31** 735.
[2] Moulton J and Smith P 1992 *Polymer* **33** 2340.
[3] Ito M, Tsuruno A, Osawa S and Tanaka K 1992 *Polymer* **29** 1161
[4] Elsenbaumer R L, Jen K Y and Oboodi R 1986 *Synth. Met.* **15** 169.
[5] Yoshino K, Nakajima S, Fujii M and Sugimoto R 1987 *Polym. Commun.* **28** 309
[6] Rughooputh S D D V, Hotta S, Heeger A J and Wudl F 1987 *J. Polym. Sci. Polym. Phys.Ed.* **25** 1071.
[7] Winokur M J, Wamsley P, Moulton J, Smith P and Heeger A J 1991 *Macromolecules* **24** 3812.
[8] Iijima S and Ichihashi T 1993 *Nature* **363** 603.
[9] Samitsu S, Shimomura T, Heike S, Hashizume T and Ito K 2008 *Macromolecules* **41** 8000.
[10] Farmer, V C, Fraser A R and Tait J M 1977 *J. Chem. Soc., Chem. Commun.* **13** 462
[11] Rusell T P 1990 *Mater. Sci. Rep.* **5** 171.
[12] Prosa T, Winokur M J, Moulton J, Smith P and Heeger A J 1992 *Macromolecules* **25** 4364.
[13] McCullough R D, Tristram-Nagle S, William S P, Lowe R D and Jayaraman M 1993 *J. Am. Chem. Soc.* **115** 4910.
[14] Inganas O, Salaneck W R, Osterholm J E and Laakso J 1988 *Synth. Met.***22** 395.
[15] Cradwick P D G, Farmer V C, Russell J D, Masson C R, Wada K and Yoshinaga N 1972 *Nat. Phys. Sci.***240** 187.
[16] Yang H, Park S, Kim D, Oh K, Magonov S, Cho K, Chang T, Bao Z and Ryuu C Y 2003 *Polym. Prepr.* **44** 33.
[17] Yang H, Shin T J, Yang L, Bao Z, Ryuu C Y, and Cho K 2004 *Polym. Prepr.* **45** 212..

Journal of Physics: Conference Series **272** (2011) 012022 doi:10.1088/1742-6596/272/1/012022

Anomalous Small-angle X-ray Scattering Study on Aggregation of a Block Copolymer in a Selective Solvent

Megumi Sakou[1], Atsuro Takechi[1], Masashi Handa[2], Yuya Shinohara[2,4], Yoshiyuki Amemiya[2,4], Hiroyasu Masunaga[3], Hiroki Ogawa[3], Naoto Yagi[3,4], Kazuo Sakurai[1,4], and Isamu Akiba[1,4]

[1]Department of Chemistry, The University of Kitakyushu
[2]Department of Advanced Materials Science, The University of Kyoto
[3]Japan Synchrotron Research Institute
[4]JST-CREST

E-mail: akiba@env.kitakyu-u.ac.jp

Abstract. Anomalous small-angle X-ray Scattering (ASAXS) analysis for poly(4-vinylphenol)-*b*-poly(4-bromostyrene) (PVPh-*b*-PBrS) in *N,N,N',N'*-tetramethyl ethylene diamine (TMEDA) solution was examined. The ASAXS measurements, executed at three different energies, provided the separation of the scattering of the Br atoms from whole system. The energy-dependent term in the SAXS profile, which is corresponding to the form factor of PBrS domain, sufficiently agree with the SAXS profile calculated for hard sphere with 41 nm radius. Therefore, the PBrS chains assemble into spherical core with 41 nm radius in TMEDA solution.

1. Introduction

Anomalous small-angle X-ray scattering (ASAXS) using absorption edge of a specific element is expected to be a powerful tool for analysis of internal structure of multiphase system because ASAXS reveals the localization of atoms or molecules in a surrounding media [1-4]. The basis of the ASAXS method is the energy-dependence of the scattering factors of the atoms, which show a remarkable change with the energy of incident X-ray in the vicinity of the absorption edge of the targeted atom [5]. Up to now, ASAXS has mainly been applied to inorganic or metallic materials because the absorption edges of the elements contained in these materials could easily be reached. On the other hand, ASAXS has hardly been used for polymer systems. For polymer systems, ASAXS using K-edge of bromine is expected to be useful because sufficient amount of Br atoms as ASAXS probe can be easily introduced into organic molecules through C-Br covalent bond and the range of energy of incident X-ray requested in ASAXS measurements near Br K-edge is available in general synchrotron light source [6]. Therefore, to achieve ASAXS method using K-edge of Br for polymer systems is of great importance for detail structural analyses of polymer materials.

Thus, in this study, we try to apply ASAXS method near the K-edge of Br atom to analysis of the internal structure of block copolymer composed of poly(4-vinylphenol) (PVPh) and poly(4-bromostyrene) (PBrS) in *N,N,N',N'*-tetramethylethylenediamine (TMEDA), which is a selective solvent and cross-linking agent through hydrogen bonds for PVPh. In this system, structural analysis for dispersed particle composed of PBrS by normal SAXS method is prevented by network of PVPh-TMEDA matrix. Therefore, ASAXS in the vicinity of K-edge of Br for this system is expected to remove the scattering from matrix and reveal the structure of PBrS domain.

2. Experimental Section

2.1. Synthesis of PVPh-*b*-PBrS and Preparation of Solution

Poly(4-*t*-butoxystyrene)-*b*-poly(4-bromostyrene) (P*t*BS-*b*-PBrS) diblock copolymer was synthesized by sequential nitroxide-mediated radical polymerization (NMRP) by using TEMPO and AIBN (Scheme 1). PtBS-*b*-PBrS was converted to poly(4-vinylstyrene)-*b*-poly(4-bromostyrene) (PVPh-*b*-PBrS) added excess of trifluoroacetyl (TFA) in THF containing excess amount of trifluoroacetic acid at room temperature. Table 1 shows molecular characteristics of PBrS and PVPh-*b*-PBrS. The resulting PVPh-*b*-PBrS was dissolved in TMEDA at 50 mg mL^{-1}. Here, TMEDA is the good solvent only for PVPh and forms cross-linking network with PVPh through hydrogen bond between tertiary amino groups and PVPh [7]. In addition, in this system, PBrS forms homogeneously dispersed spherical domains.

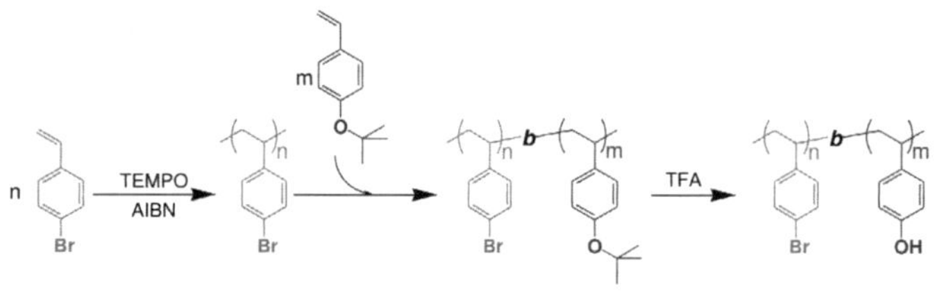

Scheme 1. Synthesis of PVPh-*b*-PBrS

Table 1. Molecular Characteristics of PBrS and PVPh-*b*-PBrS.

Sample	M_n	M_w	PDI
PBrS	$4.1x10^3$	$4.4x10^3$	1.1
PVPh-*b*-PBrS	$2.2x10^4$	$4.4x10^4$	2.0

M_n: Number-average molecular weight
M_w: Weight-average molecular weight
PDI=M_w/M_n

2.2. ASAXS measurement

Anomalous Small-angle X-ray Scattering (ASAXS) measurements were carried out at BL40B2 SPring-8, Japan. A 30 cm x 30 cm imaging plate detector (Rigaku R-AXISⅦ) was placed 4.0 m away from the sample position. To obtain the SAXS data with high S/N ratio, a bespoke vacuum sample chamber was used. The measured SAXS intensities were corrected to an absolute scale using the absolute scattering intensities of the water [8]. The exact position of the K-edge of Br in the PVPh-*b*-PBrS was determined by the measurements of X-ray transmission. The imaginary part of the anomalous scattering factor, *f"*, is proportional to the absorption coefficient and is converted to *f'* through the Kramers-Kronig (K-K) transform. Figure 1 shows *f'* and *f"* for PVPh-*b*-PBrS in TMEDA. The measurement provided 13.384 keV of the absorption edge of Br. Thus, the ASAXS measurements were performed at 13.473, 13.433, and 13.383 keV of the incident X-ray energy.

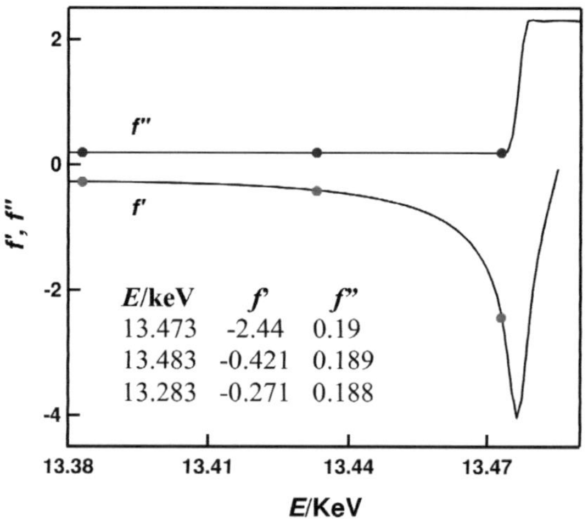

Figure 1. Anomalous dispersion corrections for Br obtained by PVPh-*b*-PBrS in TMEDA solution and the value of *f'* and *f"* at the energies used. Absorption edge is 13.483keV and especially *f'* shows drastic change in the vicinity of absorption edge.

3. Results and Discussion

The energy-dependence of SAXS intensity $I(q,E)$ can be written in the following form:

$$I(q,E)=\left\{U^2(q)+2f'(E)U(q)V(q)+\left[f'^2(E)+f''^2(E)\right]V^2(q)\right\}S(q) \quad (1)$$

where $V(q)$ is Br contribution of the scattering amplitude, $U(q)$ is the non-resonant term of the scattering amplitude, $S(q)$ is the structure factor. In dilute concentration, $S(q)$ can be regarded as 1. In equation (1), scattered intensity can be divided into three terms, a resonant term, a non-resonant term and a cross-term, and the energy dependence of the scattered factor of bromine contains anomalous dispersion coefficients $f'(E)$ and $f''(E)$. Figure 2 shows energy-dependence of SAXS profiles of PVPh-*b*-PBrS in TMEDA. SAXS intensity is decreased with increasing the energy of incident X-ray due to the energy dependence of f' of PBrS. This result suggests that ASAXS analysis can be applied to the PVPh-*b*-PBrS system. In lowest q-region, $I(q)$ shows strong q dependence. The steep slopes in the SAXS profiles should be caused by network structure composed of PVPh and TMEDA. Therefore, they should be included in $I'(q)$ in equation (1). By using equation (1), the difference of two scattering curves, measured at two different energies, is given by the following equation:

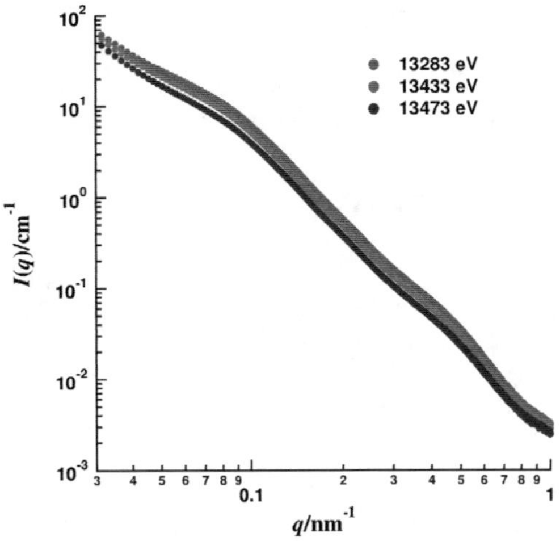

Figure 2. ASAXS measurements of PVPh-*b*-PBrS in TMEDA. *I(q)* decreased with incoming X-ray energies near the absorption edge of Br at 13483 eV.

$$\Delta I(q,E_1,E_2)=2[f'(E_1)-f'(E_2)]U(q)V(q)+[f'^2(E_1)-f'^2(E_2)+f''^2(E_1)-f''^2(E_2)]V^2(q) \quad (2)$$

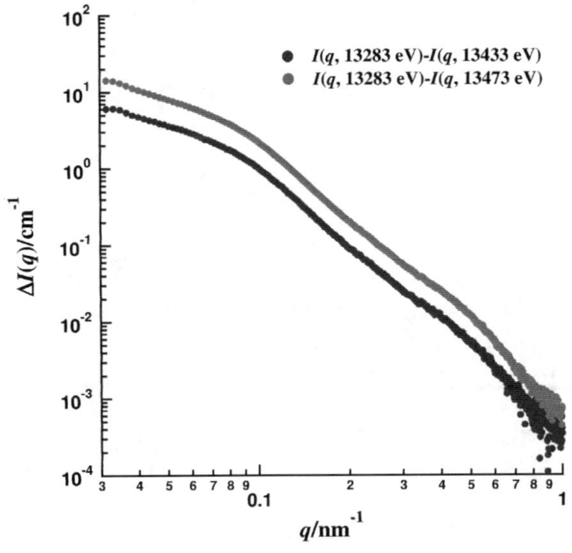

Figure 3. Separated ASAXS measurements of PVPh-*b*-PBrS in TMEDA according to equation (2).

According to equation (2), $I'(q)$ and $U^2(q)$ can be removed in the $\Delta I(q,E_1,E_2)$. Figure 3 shows $\Delta I(q,E_1,E_2)$ profiles for PVPh-*b*-PBrS. In this figure, steep slopes in the lowest q region is disappeared. Therefore, as mentioned above, the strong q dependence in the lowest q region in Figure 2 reflects the $I'(q)$ contribution in SAXS profiles. Although $I'(q)$ and $U^2(q)$ can be removed as shown in Figure 3 by equation (2), the cross-term of $U(q)$ and $V(q)$ is still remained. To analyze structure of PBrS domains, the cross-term must be removed from the SAXS profiles. The cross-term in equation (2) can be removed by further difference between $\Delta I(q,E_1,E_2)$ and $\Delta I(q,E_1,E_3)$ as the following equation.

$$V^2(q)=\frac{1}{K}\left[\frac{\Delta I(q,E_1,E_2)}{f'(q,E_1)-f'(q,E_2)}-\frac{\Delta I(q,E_1,E_3)}{f'(q,E_1)-f'(q,E_3)}\right] \quad (3)$$

$$K = f'(q,E_2) - f'(q,E_3) + \frac{f''^2(q,E_1) - f''^2(q,E_2)}{f'(q,E_1) - f'(q,E_2)} - \frac{f''^2(q,E_1) - f''^2(q,E_3)}{f'(q,E_1) - f'(q,E_3)} \quad (4)$$

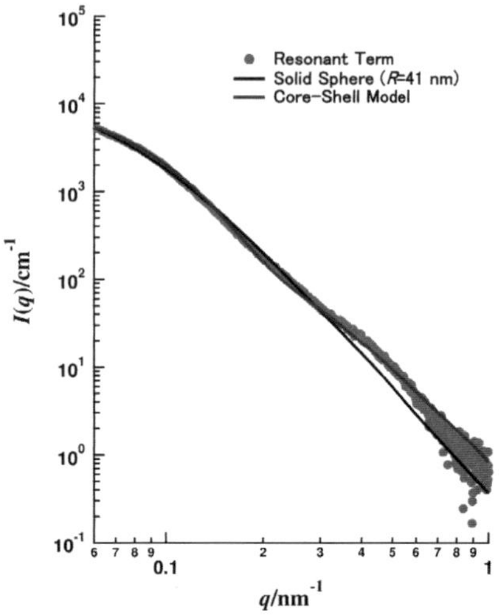

Figure 4. Resonant term obtained from ASAXS measurement for PVPh-*b*-PBrS in TMEDA (ash), and SAXS profiles calculated for hard sphere with 41 nm radius (Black). and Core-Shell model with same radius (Blue).

Figure 4 shows the $V^2(q)$ obtained from $\Delta I(q,E_1,E_2)$ and $\Delta I(q,E_1,E_3)$ by using equations (3) and (4). As mentioned above, the $V^2(q)$ is corresponding to the form factor of PBrS domain because Br atoms are included only in PBrS domains. In $q > 0.3$ nm^{-1}, $I(q)$ shows q^{-4} dependence, that means the PBrS domains have sharp interfaces. The $V^2(q)$ can be fitted with the form factor calculated for the hard sphere with 41 nm radius given in the following equation.

$$P(q) = \left[\frac{3\sin(qR) - qR\cos(qR)}{(qR)^3} \right]^2 \quad (5)$$

where R is radius of the spherical particle. Therefore, PBrS domains form spherical shape with 41 nm radius and with sharp interface against matrix. In addition, we can clearly observe bump around q=0.5 nm^{-1}. Using the core-shell model with 41 nm micelle radius and 8.5 nm core radius. In this model, electron density of shell is lower than that of core. This result should be caused by slight dissolution of TMEDA or PVPh in the outer layer of PBrS core.

4. Conclusion

ASAXS measurements near the K edge of Br for PVPh-*b*-PBrS in TMEDA reveal that PBrS domains form core-shell sphere (R=41 nm) with sharp interface in the cross-linked network of PVPh and TMEDA. Therefore, PVPh or TMEDA should partially dissolve in the outer layer of PBrS core. Thus, ASAXS analysis makes detail analysis of morphology of dispersed phase and inside of dispersed domains in a multiphase polymer system possible.

5. Acknowledgement

This study is financially supported by JST-CREST Program and the ministry of Education and Science via a grant-in-aid for IA (20550193). SAXS measurements were carried out at SPring-8 40B2 (2009B1202 and 2010A1089).

References

[1] Berg Rasmussen F, Molenbroek A M, Clausen B S, and Feidenhans R 2000 *J. Catal.* **199** 354
[2] Polizzi S, Riello P, Goerigk G, and Benedetti A 2002 *J. Synchrotron Radiat.* **9** 65
[3] Bota A, Goerigk G, Drucker T, Haubold HG, and Petro J 2002 *J. Catal.* **205** 354
[4] Dingenouts N, Patel M, Rosenfeldt S, Ponioni D, Narayanan T, and Ballauff M 2004 *Macromolecules* **37** 8152.
[5] Waseda Y *Anomalous X-ray Scattering for Materials Characteriaztion: Atomic Scale Structure Determination* 2002 Springer Berlin
[6] Varga Z, Bota A, and Goerigk G 2006 *J. Phys. Chem. B.* **110** 11029.
[7] Akiba I, and Akiyama S 1999 *Macromolecules* **32** 3741.
[8] Orthaber D, Bergmann A, and Glatter O 2000 *J. Appl. Cryst.* **33** 218.

Metalation-induced Micelle Formation of a Block Copolymer in Non-selective Solvent

Hiroyasu Masunaga[2], Ayumi Nakano[1], Kenji Yamamoto[1], Kazuya Uezu[1], Kazuo Sakurai[1,3], and Isamu Akiba[1,3*]

[1]Department of Chemistry, The University of Kitakyushu, 1-1 Hibikino, Wakamatsu, Kitakyushu 808-0135, Japan. [2]Japan Synchrotron Research Institute. [3]JST-CREST

*Corresponding author: akiba@env.kitakyu-u.ac.jp

Abstract. Metalation-induced micelle formation of polystyrene-*block*-poly[4-(2-aminoethyl amino)styrene] (PS-*b*-PDAES) with Cu^{2+} in THF, which is a good solvent for both PS and PDAES blocks, was investigated. The PS-*b*-PDAES makes dissolution of $CuCl_2$ into THF upto 1.0 x 10^{-4} mol/L (ca. 1/50 molar against diamine unit of PDAES) possible. Small-angle X-ray scattering measurements for THF solution of PS-*b*-PDAES with Cu^{2+} showed PS-*b*-PDAES forms spherical particle with 30 nm of core diameter, although PS-*b*-PDAES in THF without Cu^{2+} forms a Gaussian coil with 2.8 nm of the radius of gyration. Therefore, it was concluded that the micellar aggregation of PS-*b*-PDAES in THF by adding Cu^{2+} was induced by metalation of diamine units of PDAES and Cu^{2+}.

1. Introduction

Block copolymers composed of dissimilar blocks can form polymeric micelles with core-shell architecture in a selective solvent [1]. The formation of polymeric micelles is based on van der Walls interactions among the solventphobic blocks in block copolymers. Hence, the cores of polymeric micelles are mainly composed of solventphobic blocks in the block copolymers. The solventphobicity of selective blocks in block copolymers can be induced by external stimuli, such as pH, temperature, and pressure [2-4].

It has been well known that strong and multiple interactions, such as hydrogen bonding, electrostatic attractive interaction, and metalation with metal ion, between a polymer and the other moieties cause drastic changes in solubility of the polymer for a solvent [5-7]. Therefore, introducing such a strong interaction into one block of block copolymers should induce micelle formation of the block copolymers even in a good solvent for all blocks [8].

It has been well known that diamine derivatives can form complex with heavy metal ions, such as Cu^{2+} [9,10]. Therefore, solubility of a polymer having diamine groups for a good solvent should be drastically changed by metalation with Cu^{2+} ions. Consequently, a block copolymer containing the diamine groups in one block should form polymeric micelle having the solventphobic core composed of complex of the polymer having diamine units and Cu^{2+} via metalation between diamine units and Cu^{2+} ions in a good solvent for all block of the block copolymer under consideration. Thus, in this study, we investigated metalation-induced micelle formation of the block copolymer composed of polystyrene (PS) and poly[4-(2aminoethylamino)styrene] (PDAES) chelating with Cu^{2+} in tetrahydrofuran (THF), which is a good solvent for both PS and PDAES.

2. Experimental section

2.1. Synthesis of PS-*b*-PDAES and Preparation of Solution

The 4-(2-aminoethyl amino)styrene (DAES) is synthesized by reaction between *p*-chloromethyl styrene and 1,2-diamino ethane. The PS-*b*-PDAES was synthesized by sequential nitroxide-mediated radical polymerization (NMRP) technique using TEMPO as a stable radical. First of all, PS was synthesized by NMRP of styrene at 125 °C under N_2 atmosphere. Using the resulting PS ($M_n = 4.8 \times 10^3$, $M_w = 5.0 \times 10^3$ determined by GPC calibrated by the standard PS) as a macroinitiator, DAES was polymerized under same condition. Fig. 1 shows the structure and molecular weight of the resulting PS-*b*-PDAES. The M_n and M_w of PS-*b*-PDAES were determined by GPC calibrated by standard PS and ^{1}H-NMR.

The resulting PS-*b*-PDAES was dissolved in THF at 2.0 mg/mL (5.2×10^3 mol/L of diamine unit). For the THF solution, desired amount of $CuCl_2$ was added and dispersed with ultrasonicator.

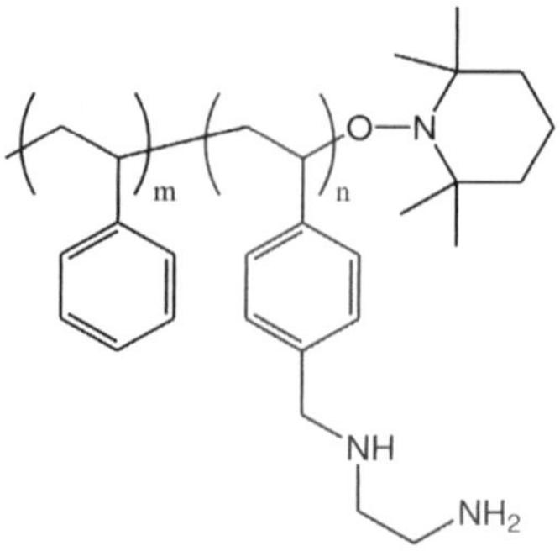

PS-*b*-PDAES
$M_n = 9.1 \times 10^3$, $M_w = 1.2 \times 10^4$

Fig. 1. Molecular structure of PS-*b*-PDAES.

2.2. Measurements

For THF solutions of PS-*b*-PDAES containing Cu^{2+}, dynamic light scattering (DLS), UV-Vis spectra, and small-angle X-ray scattering (SAXS) measurements were performed. DLS measurements were performed at 25 °C at 90° scattering angle with He-Ne gas laser (633 nm) using an Otuka Electronics DLS-7000 photometer. UV-Vis absorption measurements were performed using a JASCO V-550 UV-Vis spectrometer. SAXS measurements were carried out at BL-40B2 station of SPring-8, Japan. The wavelength of incident X-ray and exposure time were 0.10 nm and 5 min, respectively. A 30 cm x 30 cm imaging plate detector (Rigaku R-AXIS Ⅶ) was placed 1.8 m away from the sample position. To obtain the SAXS data with high S/N ratio, a bespoke vacuum sample chamber was used [11, 12]. The two-dimensional SAXS patterns obtained by the imaging plate were converted to one-dimensional $I(q)$ vs. q profiles by circular averaging, where $I(q)$ is the scattered intensity as a function of q, and q is the magnitude of scattering vector.

3. Results and discussion

Fig. 2 shows plot of absorbance at 615 nm for PS-*b*-PDAES solutions against added Cu^{2+}. When Cu^{2+} was added larger than 1.0×10^{-4} mol/L for the 2.0 ml/mL THF solution of PS-*b*-PDAES, $CuCl_2$ was not perfectly dispersed in the solution. Therefore, for the solution with $[Cu^{2+}]>1.0$ mol/L, UV-Vis spectra were measured after filtration. Absorbance is linearly increasing with increasing Cu^{2+} up to 1.0×10^{-4} mol/L for 2.0 mg/mL THF solution of PS-*b*-PDAES (5.2×10^{-3} mol/L of diamine unit). Therefore the saturation concentration of Cu^{2+} in

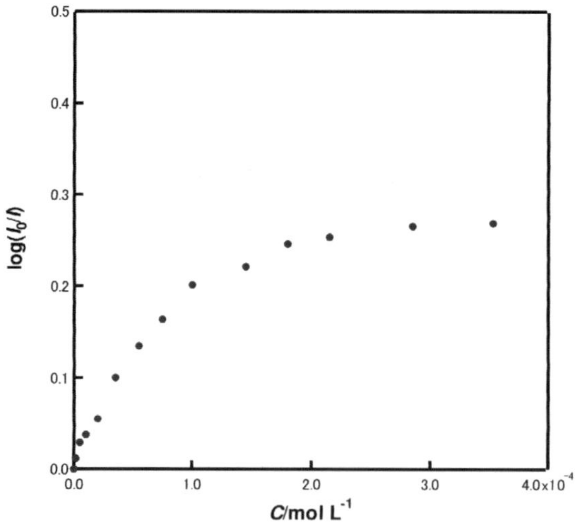

Fig. 2. Plot of absorbance of visible light at 615 nm against added amount of Cu^{2+} for THF solution of PS-*b*-PDAES at 2.0 ml/mL.

Journal of Physics: Conference Series **272** (2011) 012023 doi:10.1088/1742-6596/272/1/012023

THF containing 2.0 mg/mL of PS-*b*-PDAES is estimated as 1.0x10^{-4} mol/L although Cu^{2+} is scarcely dissolved in THF without PS-*b*-PDAES. This result indicates that the dissolution of Cu^{2+} into THF solution is induced by metalation between diamine-unit of PDAES and Cu^{2+}. Consequently, it should be expected that micelle formation of PS-*b*-PDAES is induced by the metalation in THF solution. To confirm the micelle formation of PS-*b*-PDAES in THF, SAXS measurements are carried out.

Fig. 3 shows SAXS profiles from THF solutions of PS-*b*-PDAES at 2.0 mg/mL containing 1.0 x 10^{-4} mol/L of Cu^{2+} and without Cu^{2+}. $I(q)$ of PS-*b*-PDAES/Cu^{2+} shows q^{-4} dependence, although that of I(q) of PS-*b*-PDAES without shows q^{-2} dependence. The q^{-4} and q^{-2} dependences of $I(q)$ characterize the scattering from sharp interface (Porod's low) and Gaussian coil, respectively. Therefore, PS-*b*-PDAES in THF without Cu^{2+} behaves like a homopolymer in a good solvent. The SAXS profile of PS-*b*-PDAES in THF without Cu^{2+} is well fitted with the scattering profile calculated for Gaussian coil of 2.8 nm of radius of gyration (R_g). On the other hand, the SAXS profile of PS-*b*-PDAES in THF containing Cu^{2+} shows good agreement with theoretical scattering curve calculated for solid sphere with 15.0 nm radius (R) using the following equation.

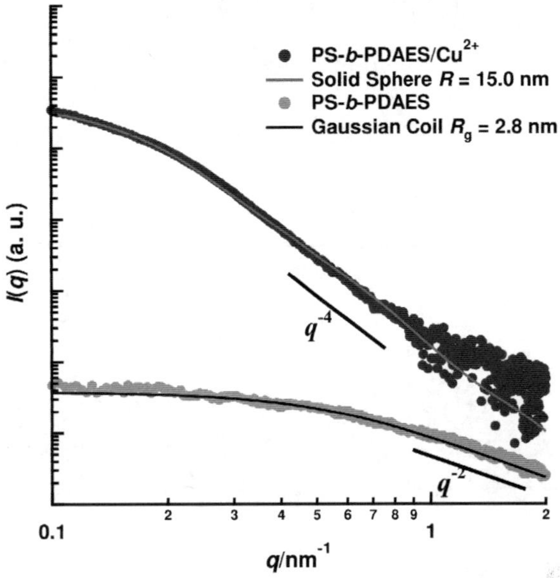

Fig. 3. SAXS profiles from THF solutions of PS-*b*-PDAES at 2.0 mg/mL containing 1.0 x 10^{-4} mol/L of Cu^{2+} (blue) and without Cu^{2+} (gray). The solid lines are the theoretical scattering curves calculated for Gaussian coil of 2.8 nm R_g and solid sphere with 15.0 nm radius (red), respectively (black).

$$I(q) \propto \left[\frac{3\{\sin(qR) - qR\cos(qR)\}}{(qR)^3} \right]^2 \tag{1}$$

Taking into account the polydispersity as a Gaussian distribution and using standard deviation σ= 4 (σ/R=0.27), the fitting analyses have been successfully done. In this case, the polymeric micelle should form core-shell structure having solventphobic core composed of complex of PDAES with Cu^{2+} and shell of the mixture of PS and THF. When the polymeric micelle takes core-shell structure, the SAXS profile of the micelle should be expressed by the profile calculated for the core-shell model given by the following equation [13].

$$\begin{aligned} I(q) &\propto \left[(\rho_c - \rho_s)v_c f_c(q) + (\rho_s - \rho_0)v_s f_s(q)\right]^2 \\ f_i(q) &= \frac{3\{\sin(qR_i) - qR_i\cos(qR_i)\}}{(qR_i)^3} \\ i &= c, s \end{aligned} \tag{2}$$

Here, R_c and R_s are radii of core and overall micelle; ρ_c, ρ_s, and ρ_0 are electron densities of core, shell, and solvent, respectively. In this case, ρ_c-ρ_s should be much higher than ρ_s-ρ_0. Therefore, the second term of equation (2) can be ignored. Thus, it is considered that the SAXS profile from PS-*b*-PDAES/Cu^{2+} shows good agreement with the theoretical scattering profile calculated for simple hard sphere model. Consequently, the PS-*b*-PDAES/Cu^{2+} micelle should have the solventphobic PDAES/Cu^{2+} core with 15 nm radius. According to dynamic light scattering measurements, radius of overall PS-*b*-PDAES/Cu^{2+} micelle is estimated to 26 nm. Using these results, we can derive the conclusion schematically represented in Fig. 4for mechanism of formation of PS-*b*-PDAES micelle in THF solution by metalation of diamine units of PDAES with Cu^{2+}.

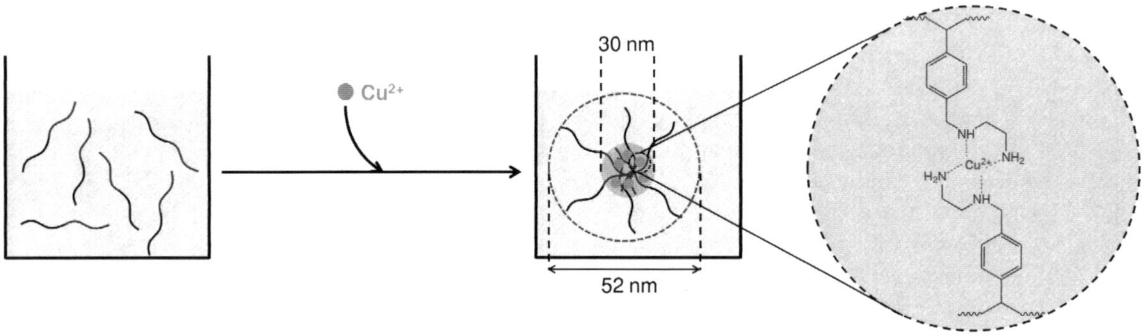

Fig. 4. Schematic representation of mechanism of micelle formation of PS-*b*-PDAES with Cu^{2+} in THF solution.

4. Acknowledgement

This study is financially supported by JST-CREST Program, Technology Transfer Program of JST, and the ministry of Education and Science via a grant-in-aid for IA (20550193). SAXS measurements were carried out at SPring-8 40B2 (2009B1202 and 2010A1089).

References

[1] Zhang M, and Eisenberg A 1995 *Science* **268** 1728.
[2] Hu Y, Kramer MC, Boudreaux CJ, and McCormic CL 1995 *Macromolecules* **28** 7100.
[3] Ringsdorf H, Simon J, and Winnik FM 1992 *Macromolecules* **25** 7306.
[4] Zhou S, and Chu B 1998 *Macromolecules* **31** 5300.
[5] Akiba I, Seki T, and Akiyama S 2002 *e-Polym.* 8.
[6] Akiba I, and Akiyama S 2001 *J. Macromol. Sci. Phys.* **B40** 157.
[7] Wang LF, Pearce EM, and Kwei TK 1991 *J. Polym. Sci. Polym. Phys. Ed.* **29** 619.
[8] Yoshida E, and Kunugi S 2002 *Macromolecules* **35** 6669.
[9] Tokuyama H, Fujioka M,and Sakohara S 2005 *J. Chem. Eng. Jpn.* **38**, 633.
[10] Nishimura T, Masunaga H, Ogawa H, Akiba I, and Sakurai K 2010 *Chem. Lett.* **39**, 686.
[11] Sakuragi M, Kusuki S, Hamada E, Masunaga H, Ogawa H, Akiba I, and Sakurai K 2009 *J. Phys. Conf. Ser.* **184** 012008.
[12] Akiba I, Terada N, Hashida S, Sakurai K, Sato T, Shiraishi K, Yokoyama M, Masunaga H, Ogawa H, Ito K, and Yagi N 2010 *Langmuire* **26** 7544.
[13] Wu G, Chu B, and Schneider DK 1994 *J. Phys. Chem.* **98** 12018.

Journal of Physics: Conference Series **272** (2011) 012024 doi:10.1088/1742-6596/272/1/012024

Crystallization of isotactic polypropylene from mesomorphic phase: a constant heating rate study

H Asakawa[1], K Nishida[1*], G Matsuba[2], T Kanaya[1], and H Ogawa[3]

[1] Institute for Chemical Research, Kyoto University, Uji, Kyoto, 611-0011, Japan
[2] Department of Polymer Science and Engineering, Yamagata University, Yonezawa, Yamagata, 992-8510, Japan
[3] Japan Synchrotron Radiation Research Institute (JASRI), Kouto, Sayo-cho, Sayo-gun, Hyogo, 679-5198, Japan

[*]E-mail: knishida@scl.kyoto-u.ac.jp

Abstract. We have studied crystallization behaviour of isotactic polypropylene (iPP) from mesomorphic phase in structural point of view. Time-resolved wide-angle X-ray diffraction (WAXD) measurements during a heating process have been performed using a synchrotron radiation (SR) X-ray beam line at SPring-8, Japan. The heating process was so programmed to reproduce a thermal trace of differential scanning calorimetry (DSC) with a constant heating rate (10 °C/min) in order to compare the structural change with thermal behaviour. SR-WAXD sensitively detected the crystallization behaviour and we have obtained fractions of alpha-crystal, mesomorphic phase and amorphous phase as a function of temperature by analysing the data. The results showed that the crystallization from mesomorphic phase proceeds in between 60 and 120 °C (meso-alpha transition). During this process, the crystallization from amorphous hardly takes place. The crystalline fraction shows almost constant in between 120 and 140 °C; meanwhile, the mesomorphic fraction still decreases above 120 °C. The crystalline fraction starts to decrease above 140 °C and the most extensively decreases at around 165 °C (melting point). We have also determined the energy level of the mesomorphic phase (meta-stable state) relative to that of alpha-crystal (stable state), considering the balance among the fractions of alpha-crystal, mesomorphic phase and amorphous.

1. Introduction

From a standpoint of thermal process, crystallization of bulk polymer is classified into mainly two categories; one is cooling from molten state and another is heating from "pre-quenched state". In both processes usually growth of spherulites in micrometer scale are observed, though the latter process yields more or less higher nucleus density than the former one. In those polymers "pre-quenched state" is mostly equivalent to "glassy state". However, this is not the case for isotactic polypropylene (iPP). In the heating process of pre-quenched iPP, the growth of spherulites is never observed and the macroscopic appearance is almost transparent up to the melting temperature T_m. The absence of the spherulites and the transparency not necessarily imply that the pre-quenched iPP does not crystallize. The pre-quenched iPP indeed crystallizes in the heating process [1], but the higher order structure is quite different from one that is crystallized from molten state [2]. The specific behaviour of the crystallization of iPP from the pre-quenched state is considered to be due to the presence of mesomorphic phase. IPP has three well-established polymorphisms; such as alpha (monoclinic), beta (hexagonal) and gamma (orthorhombic) forms [3], however, the mesomorphic phase does not correspond to any polymorphisms. The structure of mesomorphic phase of iPP shows intermediate one between the crystal and amorphous [4]. The mesomorphic phase of iPP is obtained by rapidly quenching molten iPP into low temperature, empirically 0 °C. The mesomorphic iPP lasts stably at room temperature (RT) without transforming into crystal although the nominal glass transition temperature T_g is below RT [5]. Formation of the mesomorphic phase of iPP is exothermic process as was demonstrated by a high-speed differential scanning calorimetry [6, 7]. Considering these characteristics, the mesomorphic phase of iPP is considered a meta-stable state whose energy level is very low comparable to that of crystal.

Differential scanning calorimetry (DSC) is usually a versatile method to examine various transitions among states in polymer materials. However, this method is a rather insensitive to investigate the transition of iPP from mesomorphic phase to crystal, since the majority of the exotherm was completed during the formation of the mesomorphic phase and the remaining exothermic enthalpy is considered to be small as was mentioned in the preceding paragraph. To make the matter more complex, other transitions such as mesomorphic to amorphous and/or amorphous to crystal may co-occur during heating, whose enthalpy are larger than that of the mesomorphic to crystal. Alternatively, the transition should be investigated in structural point of view. In this respect, wide-angle X-ray diffraction (WAXD) measurement is a sensitive method to detect the structural change whether or not the transition is accompanied by exo- or endotherm. Recent developments of high-intensity synchrotron radiation (SR) source and sensitive time-resolved detector system enable in-situ WAXD measurements in the similar thermal trace of DSC measurement. In this study, such a SR-WAXD technique has been applied to investigate crystallization of iPP from mesomorphic phase. Fractions of alpha-crystal, mesomorphic phase and amorphous phase as a function of temperature have been obtained by a component analysis of WAXD profile. The derivative of the crystalline fraction as a function of temperature clearly shows crystallization behaviour of iPP from mesomorphic phase, which backs up the above-mentioned disadvantage of DSC method.

Lastly, combining the structural information by SR-WAXD and DSC measurements, the energy level of mesomorphic phase has been estimated. Although rough estimation of the energy level of mesomorphic phase is possible by assuming the original fraction of mesomorphic phase transforms into the same fraction of alpha-crystal [8, 9], in reality the balance among the fractions of alpha-crystal, mesomorphic phase and amorphous phase as a function of temperature should be taken into consideration for the more precise estimation.

2. Experimental

2.1. Materials and sample preparations

The iPP material has a weight-average molecular weight M_w = 353000, a polydispersity of M_w / M_n = 4.4, and a degree of isotacticity (a meso pentad value) *mmmm* = 0.98, which was supplied from Idemitsu Unitech Co., Ltd. This isotacticity is higher enough than the critical value (0.68) of isotacticity to form mesomorphic phase [10]. A mesomoprhic phase thin film was obtained by quenching molten iPP to 0 °C by dipping into ice-water. The obtained film was macroscopically transparent and showed two broad peaks in WAXD, which are the characteristics of mesomorphic iPP.

2.2. Measurements

DSC measurements were carried out with a Perkin-Elmer Diamond DSC at a heating rate of 10 °C/min. WAXD measurements were carried out using the beam line BL40B2 in the SR facility, SPring-8, Hyogo, Japan. Wavelength of the incident X-ray was 1.0 Å. The detector used was CMOS flat panel detector from Hamamatsu Photonics K.K. The exposure time was selected to be 0.5 s. A Mettler FP82HT hot stage was used to heat the specimens for WAXD measurements at the same heating rate of the DSC measurement. In both DSC and WAXD measurements, thin films were tightly piled up to the same sample thickness, 0.7 mm, to apply similar thermal conduction within the samples.

3. Results and discussion

3.1. Differential scanning calorimetry (DSC)

Figure 1 shows a DSC curve for a mesomorphic iPP during a heating process. The curve shows a shoulder at around 40 - 50 °C. We assume this shoulder is due to a glass transition like phenomenon of the mesomorphic phase of iPP, though there still is discussion for the origin [11]. It will be discussed elsewhere in this regard. An exothermal shallow peak is observed at around 100 °C. As will be turned out in the WAXD analysis, this exothermal shallow peak is due to the crystallization from the mesomorphic phase. Endotherm starts at around 130 °C and a distinct endothermic peak is observed at around 165 °C. It is unquestionable to assign the distinct endothermic peak to the melting of the crystal.

Integrated areas of these exo- and endothermic peaks do not balance out. This is mainly because that the exothermic enthalpy when the mesomorphic phase transforms into crystal is small, as was mentioned in the introduction. However, it lacks proof to claim these explanations only with the DSC data. This will be discussed in the following section.

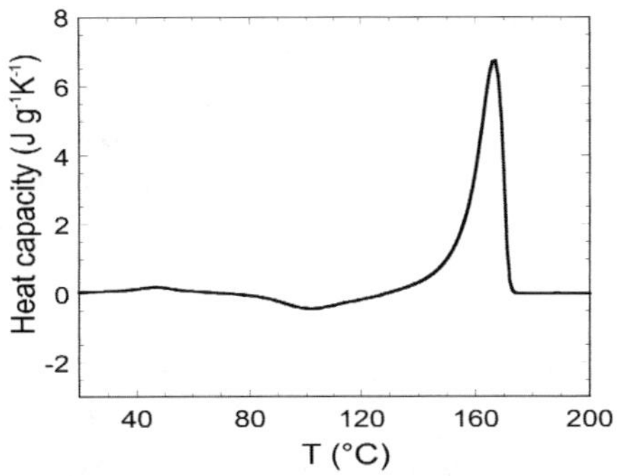

Figure 1. Observed DSC curve for mesomorphic iPP during heating process.

3.2. Wide-angle X-ray diffraction (WAXD)

Figure 2 shows in-situ WAXD profiles during a heating process from 30 to 170 °C. Broad peaks appearing at scattering vectors around q = 10.5 and 14.9 nm^{-1} for the profiles of low temperatures are characteristics of the mesomorphic iPP. Sharp peaks appearing at q = 9.7, 11.5, 12.8 and 15.1 nm^{-1} for the profiles of temperatures between 90 and 160 °C are reflections from alpha-crystal. Single broad peak appearing at q = 10 nm^{-1} for the profiles at 170 °C is the so-called amorphous halo.

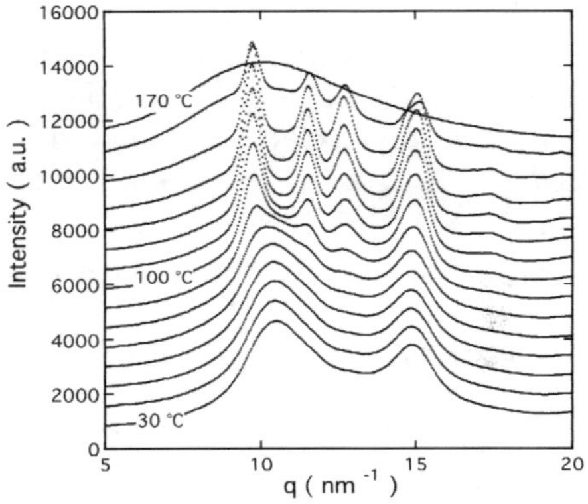

Figure 2. WAXD profiles during heating process from 30 to 170 °C displayed at every 10 °C from bottom to top. The profiles were shifted vertically for clarification.

In order to discuss the crystallization behaviour quantitatively, component analysis for WAXD profiles was carried out assuming each profile consists of three components such as alpha-crystal, mesomorphic, and amorphous, according to the previously reported method [1]. Dividing the integrated intensity of each component by the integrated intensity of observed (total) intensity for each profile, fractions of alpha-crystal Φ_{cryst}, mesomorphic Φ_{meso} and amorphous Φ_{amorph} as a function of temperature during heating process were obtained, as shown in figure 3. The crystallization from mesomorphic phase proceeds in between 60 and 120 °C (meso-alpha transition); this behaviour is the most conspicuous at around 100 °C. During this process, the crystallization from amorphous hardly takes place. The crystalline fraction shows almost constant in between 120 and 140 °C. The maximum crystalline fraction (crystallinity) at this temperature is in disagreement with the original mesomorphic fraction at RT (figure3), at

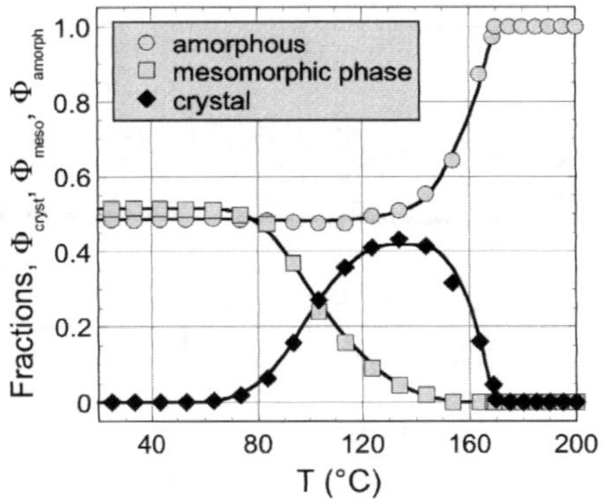

Figure 3. Fractions of alpha-crystal, mesomorphic and amorphous as a function of temperature during heating process.

least in the present experimental condition. A possible reason for this disagreement is that the heating rate of 10 °C/min is not slow enough to let the mesomorphic phase transforms into crystal to the full extent. The crystalline fraction starts to decrease above 140 °C and the most extensively decreases at around 165 °C. Temperature derivative of alpha-crystal fraction $-d\Phi_{cryst}/dT$ more clearly shows crystallization behaviour from mesomorphic phase (figure 4). The integrated areas for the crystallization peak at around 100°C and that for the melting peak at around 165°C are equivalent in principle. With the support of this results, the shallow peak at around 100°C observed in DSC (figure 1) has definitely turned out to be due to the crystallization.

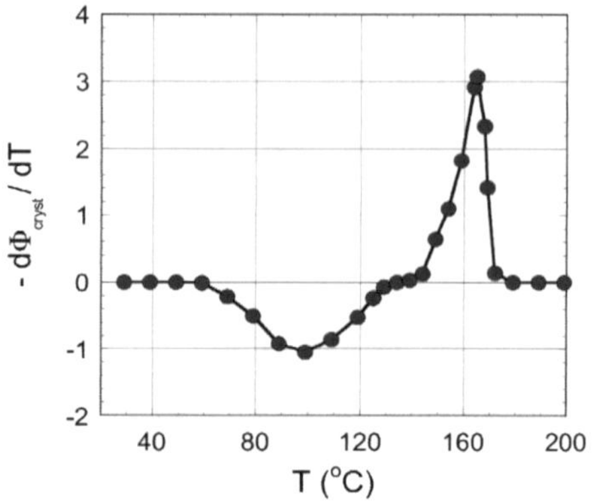

Figure 4. Temperature derivative of alpha-crystal fraction as a function of temperature.

3.3. Energy level of mesomorphic phase

Figure 5 shows schematic diagram of energy levels of alpha-crystal and mesomorphic phase relative to amorphous. Assuming the melting enthalpy of mesomorphic phase (ΔH_{meso}) being x times of that of alpha-crystal (ΔH_{cryst}), namely $\Delta H_{meso} = x\ \Delta H_{cryst}$, the total enthalpy relative to amorphous $\Delta H(T)$ is given by the following equation;

$$\Delta H(T) = \Delta H_{cryst}\Phi_{cryst}(T) + x\ \Delta H_{cryst}\Phi_{meso}(T). \qquad (1)$$

Temperature derivative of $\Delta H(T)$ gives expected value of heat balance. As a reliable value of ΔH_{cryst} was already reported [12], unknown parameter is only x. Figure 6 shows $-d(\Delta H)/dT$ plotted for several values of x. The case when $x = 0.75$ reproduces the experimental DSC well. This value is significantly smaller than the previously reported values (= 0.97 [8] and 0.93 [9]). It is essential to consider the balance among the fractions of alpha-crystal, mesomorphic phase and amorphous phase to obtain the reliable value, otherwise we end up overestimation of the value.

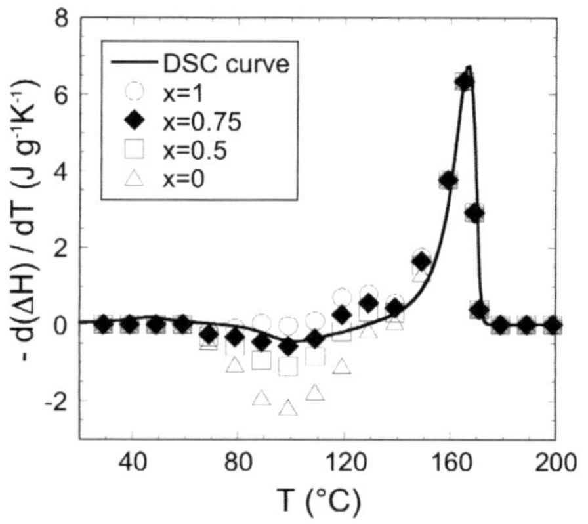

Figure 5. Temperature derivative of total enthalpy as a function of temperature for several numerical parameter x (see text).

References

[1] Konishi T, Nishida K and Kanaya T 2006 *Macromorecules* **39** 8035
[2] Nishida K, Konishi T, Kanaya T and Kaji K 2004 *Polymer* **45** 1433
[3] Turner-Jones A, Aizlewood J M, and Beckett D R 1964 *Makromol. Chem.* **75** 134
[4] Miller R 1960 *Polymer* **1** 135
[5] Miyamoto Y, Fukao K, Yoshida T, Tsurutani N and Miyaji H *J. Phys. Soc. Jpn.* 2000 **69** 1735
[6] Wu Z Q, Dann V L, Cheng S Z D and Wunderlich B 1988 *J. Thermal Anal.* **34** 105
[7] Gradys A, Sajkiewicz P, Minakov A A, Adamovsky A, Schick C, Hashimoto T and Saijo K 2005 *Materials Sci. Eng. A* **413–414** 442
[8] Cao J and Sbrrarski I 2006 *Polymer* **47** 27
[9] Grebowicz J, Lau S –F and Wunderlich B 1984 *J. Polym. Sci., Polym. Symp.* **71** 19
[10] Konishi T, Nishida K, Kanaya T and Kaji K 2005 *Macromorecules* **38** 8749
[11] Mileva D, Androsch R, Zhuravlev and Schick C 2009 *Macromolecules* **42** 7275
[12] Bu H, Cheng S Z D and Wunderlich B 1988 *Makromol. Chem. Rapid. Commun.* **9** 75

Journal of Physics: Conference Series **272** (2011) 012025 doi:10.1088/1742-6596/272/1/012025

Penetration route of functional molecules in stratum corneum studied by time-resolved small- and wide-angle x-ray diffraction

Ichiro Hatta[1], Hiromitsu Nakazawa[2], Yasuko Obata[3], Noboru Ohta[1], Katsuaki Inoue[4], Naoto Yagi[1]

[1]Japan Synchrotron Radiation Research Institute/SPring-8, Japan, [2]Department of Physics, Kwansei Gakuin University, Japan, [3]Department of Pharmaceutics, Hoshi University, Japan, [4]Diamond Light Source, U.K.

E-mail: hatta@spring8.or.jp

Abstract. We studied effects of functional molecules on corneocytes in stratum corneum using time-resolved small- and wide-angle x-ray diffraction after applying a functional molecule. From these results it was revealed that in the stratum corneum a typical hydrophilic molecule, ethanol, penetrates via the transcellular route and on the other hand a typical hydrophobic molecule, *d*-limonene, penetrates via the intercellular route.

1. Introduction

The outermost layer of skin, stratum corneum (SC), is composed of corneocytes and intercellular lipid matrix. The structure of the SC plays an important role in the barrier function [1,2]. The intercellular lipids in the SC form lamellar structures. In the lamellar repeating direction the long and the short lamellar structures which have the repeat distances of 13.5 nm and 6 nm, respectively, have been detected by the small-angle x-ray diffraction (SAXD) and in the orthogonal direction the lattice spacings of the hydrocarbon-chain packing structure, 0.41 nm and 0.37 nm, have been observed by the wide-angle x-ray diffraction (WAXD). Recently it has been proposed that in the intercellular lipid matrix there are two domains: A domain with hydrophobic character consists of the long lamellar structure with hexagonal hydrocarbon-chain packing in which the lattice spacing is 0.41 nm and another domain with hydrophilic character consists of the short lamellar structure having water layers with orthorhombic hydrocarbon-chain packing in which the lattice spacing is 0.41 nm and 0.37 nm [3]. Furthermore, it has been pointed out that very diffuse peaks with the lattice spacing of 1 nm and 0.46 nm observed by the x-ray diffraction are related to the structure of soft keratin in the corneocytes in the SC [2]. The SAXD and WAXD measurements on the SC are a powerful tool since the molecular arrangements resulting from the structural modification in the SC can be measured when functional molecules such as penetration enhancers, cosmetics, etc. are applied to the SC. In general there exist two potential pathways; one is an intercellular route in which the penetration of the functional molecule occurs via the intercellular lipid matrix lying between the corneocytes and the other is a transcellular route in which the penetration takes place across both the corneocytes and the intercellular lipid matrix. However, whether the former is the fundamental route as claimed by the majority opinion or the both play a complementary role as pointed out by some researchers is still controversial. Therefore, it is highly important to make clear the structural modification at the molecular level caused by application of the functional molecules to the SC.

Here we investigated effects of hydrophilic or hydrophobic molecule on the SC. As a typical hydrophilic or hydrophobic molecule we applied ethanol [4,5] or *d*-limonene [6-9], respectively. For this purpose we have developed an x-ray diffraction method which can perform the SAXD and WAXD measurements successively for a single SC sample as a function of time after applying the functional molecules [10]. From these measurements we are able to elucidate the possible penetration routes of a hydrophilic or hydrophobic molecule.

It is expected that incorporation of a hydrophilic molecule into the corneocytes modifies the structure of the soft keratin but that of a hydrophobic molecule does not. Thus, the structure of the soft keratin in the corneocytes is crucial in the study on the penetration route in the SC. After the application of hydrophilic or hydrophobic molecule, we focused our attention to the detection of the changes in the x-ray diffraction profile with about 1 nm spacing where a characteristic broad peak associated with the soft keratin takes place [2].

2. Experimental

Eight-week-old male hairless mice (HR-1, Hoshino, Japan) were used. The sample preparation has been described in the previous paper [10]. The small- and wide-angle x-ray experiments were performed at BL40B2 of SPring-8 (Japan). The experimental procedure was mentioned also in the previous paper [10]. The diffraction profile for the entire angular range was obtained with an imaging plate every 210 s. The exposure time was 30 s. Through the figures of the paper, the profile was shown every 210 s successively from red to blue curves. A sample cell developed by us has been described in the previous paper [10]. A stratum corneum sample was embedded in a central hollow with cylindrical shape surrounded by a solution-permeable support. The front and the rear surfaces of the hollow were sealed by a pair of thin-polyimide-film windows with a thickness of 7 μm. In the present study the sample cell was filled by a solution at the beginning of the measurement. As a hydrophilic molecule, we used ethanol (reagent grade, Wako Pure Chemical Industries, Ltd., Osaka) and as a hydrophobic molecule, *d*-limonene (reagent grade, Tokyo Chemical Industries, Co. Ltd., Tokyo).

3. Results and Discussion

To describe the overall tendencies, the successive changes of the x-ray diffraction profile for the whole angular range are shown in Figs. 1a and b. Profiles after applying ethanol and *d*-limonene to the SC sample are shown together with the diffraction profiles of pure ethanol and pure *d*-limonene, respectively. The abscissa is $S=(2/\lambda)\sin(2\theta/2)$, where λ (=0.1 nm) is the wavelength of x-ray and 2θ is the scattering angle. The apparent peak at 0.6 nm^{-1} is due to the scattering from the polyimide-film windows of the sample cell. The profile of pure ethanol exhibits extremely broad peaks near S=1.25 nm^{-1} and S=2.4 nm^{-1}. The profile of pure *d*-limonene exhibits an extremely broad peak near S=1.9 nm^{-1}. It is clearly seen in Fig. 1a and b that, by applying ethanol and *d*-limonene to the SC, the broad peaks originating from these molecules increased gradually. These results indicate the uptake of these molecules in the SC. We have already analyzed and discussed the behavior in the small angle range of $0.0<S(nm^{-1})<0.4$ and in the wide angle range of $2.0<S(nm^{-1})<3.0$ in the previous paper [10]. In this paper, since a broad peak of the soft keratin in the corneocytes appears near 1 nm^{-1} (that is, the spacing of 1 nm), we focused our attention to the range of $0.5<S(nm^{-1})<1.5$ and to comparing the changes caused by ethanol and *d*-limonene. To show the minute changes in the diffraction profiles we calculated the intensity difference $\Delta I(S,t)$ given by

$$\Delta I(S,t) = I(S,t) - I(S,0)\,, \qquad (1)$$

where $I(S,t)$ is the intensity at time t and $I(S,0)$ is the intensity just after applying the functional molecule to the SC. In Fig. 2a and b, the intensity difference is shown for ethanol and *d*-limonene in the range of $0.5<S(nm^{-1})<1.5$, respectively. With *d*-limonene, the intensity did not change near 1 nm^{-1} except the increase in the range larger than 1 nm^{-1} as a result of the uptake of *d*-limonene in the SC.

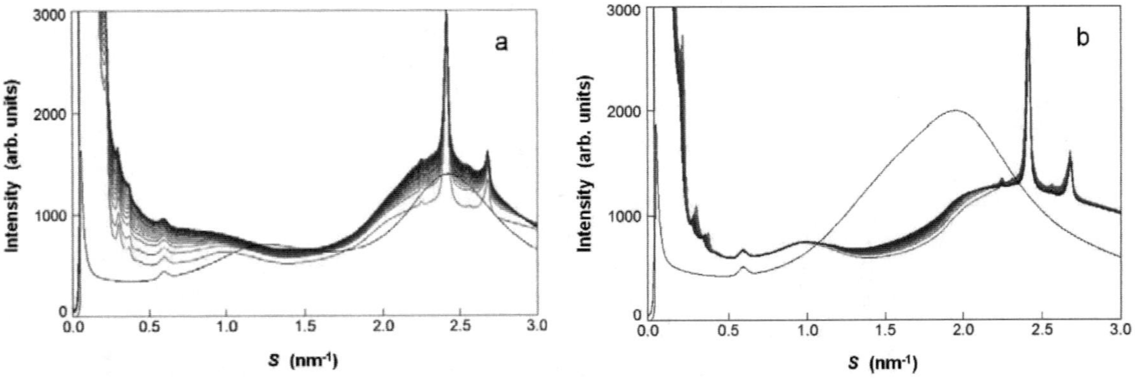

Figure 1a and b. a exhibits the whole x-ray diffraction profile changes with time on applying ethanol to the stratum corneum (from red to blue curves, see Sec. Exp.) with the profile for pure ethanol (black curve). **b** indicates the whole x-ray diffraction profile changes with time on applying *d*-limonene to the stratum corneum (from red to blue curves, see Sec. Exp.) with the profile for pure *d*-limonene (black curve).

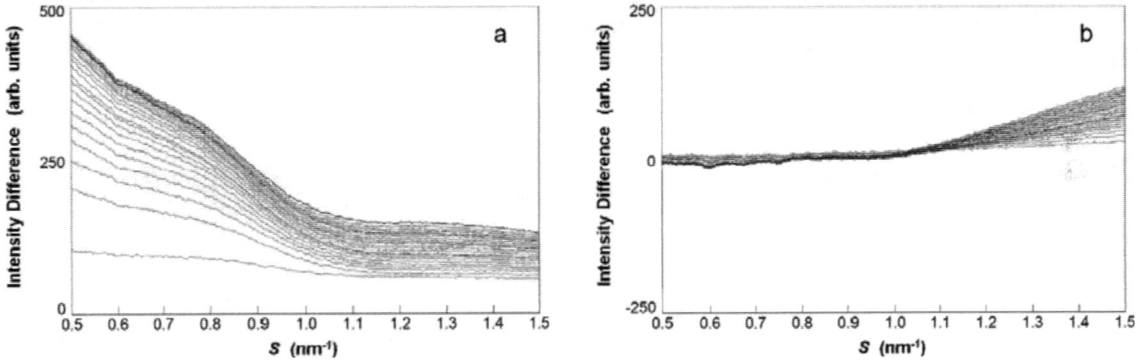

Figure 2a and b. a exhibits the successive change in the profiles of the intensity difference near 1 nm^{-1} on applying ethanol to the stratum corneum (from red to blue curves, see Sec. Exp.) [10]. **b** indicates the successive change in the profiles of the intensity difference near 1 nm^{-1} on applying *d*-limonene to the stratum corneum (from red to blue curves, see Sec. Exp.).

On the other hand, with ethanol, the intensity increased over the entire angular range. The increment was bigger in the smaller angular range. However, in superposition on this behavior the decrease of the intensity occurred near 1 nm^{-1} clearly where the broad peak of the soft keratin lay and additionally the increase of the very broad peak near 1.25 nm^{-1} occurred as a result of the uptake of ethanol in the SC. Furthermore the behavior in the smaller angular range was distinctly different between ethanol and *d*-limonene as seen in Fig. 3a and b, respectively, in which a peak at about 0.07 nm^{-1} corresponds to the first order diffraction of the long lamellar structure. The behavior of this reflection after application of ethanol or *d*-limonene has been discussed in the previous paper in detail [10]. In addition, the distinct increase of the intensity in the smaller angle region occurred on applying ethanol, but no significant change took place on applying *d*-limonene.

Based upon the present results, we can point out clearly that ethanol affects the structure of the soft keratin in the corneocytes, while *d*-limonene does not. Therefore, ethanol is incorporated into the corneocytes and then the x-ray diffraction intensity increases in the smaller angular range. Finally by referring to the results obtained in the previous paper [10] we can propose two penetration routes: One is the transcellular route as observed in the application of hydrophilic ethanol and the other is the intercellular route as observed in the application of hydrophobic *d*-limonene. Then, not only the well-

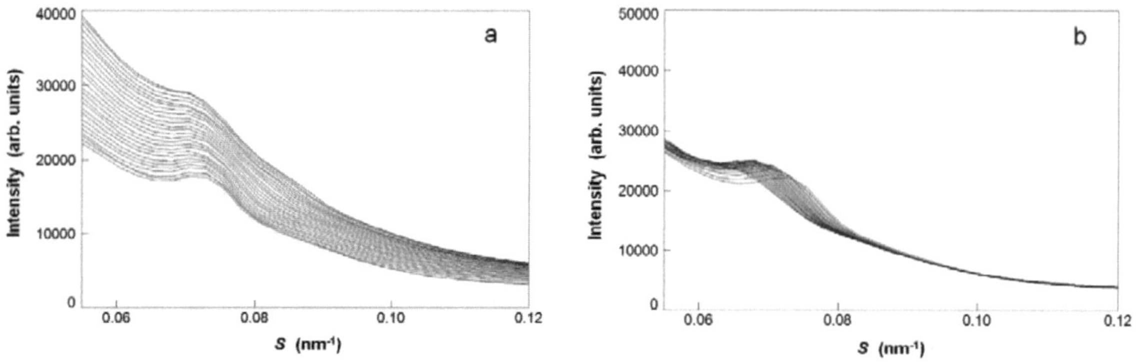

Figure 3a and b. a and **b** exhibit the time course of the profiles in the small angle region after applying ethanol and *d*-limonene to the stratum corneum, respectively (from red to blue curves, see Sec. Exp.).

known intercellular route that is schematically shown as

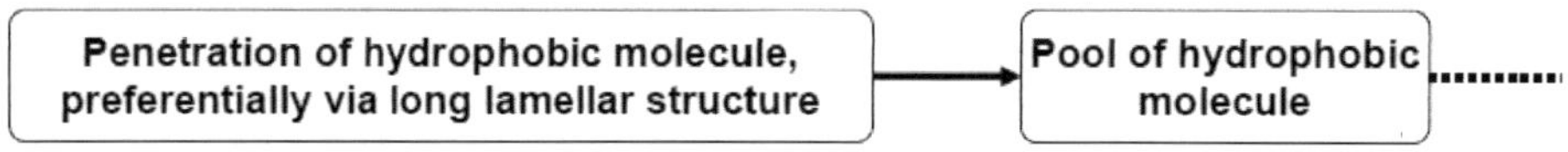

but also the transcellular route that is schematically shown as

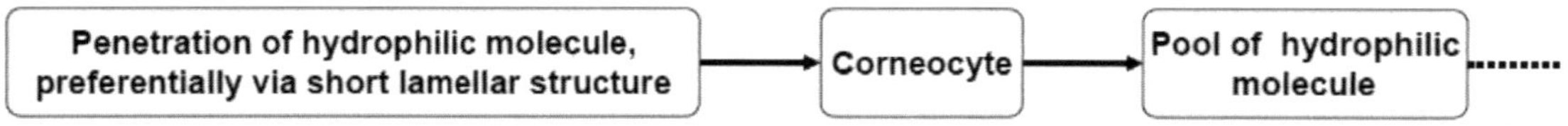

plays an indispensable role in the penetration of the functional molecules through the SC.

Acknowledgments
This work was supported by Grant-in-Aids for Scientific Research (C) (18540411) to IH and for Young Scientists (B) (1979085) to HN from the Ministry of Education, Culture, Sports, Science and Technology. The synchrotron radiation experiments were performed at the SPring-8 with the approval of the Proposal Review Committee (proposed no. 2006A1074, 2008A1458, 2009A1779 and 2010A1482).

References
[1] White, S. H., Mirejovsky, D., King, G. I., 1988. Biochemistry **27**, 3725.
[2] Bouwstra, J. A., Gooris, G. S., van der Spek, J. A., Lavrijsen, S., Bras, W., 1994. Biochim. Biophys. Acta **1212**, 183.
[3] Hatta, I., Ohta, N., Inoue, K., Yagi, N., 2006. Biochim. Biophys. Acta **1758**, 1830.
[4] Kurihara-Bergstrom, T., Knutson, K., DeNoble, L. J., Goates, C. Y., 1990. Pharm. Res. **7**, 762.
[5] Kai, T., Mak, V. H. W., Potts, R. O., Guy, R. H., 1990. J. Controlled Release **12**, 103.
[6] Okabe, H., Takayama, K., Ogura, A., Nagai, T., 1988. Drug Design & Delivery **4**, 313.
[7] Cornwell, P. A., Barry, B. W., Stoddart, C. P., Bouwstra, J. A., 1994. J. Pharm. Pharmacol. **46**, 938.
[8] Cornwell, P. A., Barry, B. W., Bouwstra, J. A., Gooris, G. S., 1996. Intern. J. Pharm. **127**, 9.
[9] Williams, A. C., Barry, B. W., 2004. Advanced Drug Delivery Reviews **56**, 603.
[10] Hatta, I., Nakazawa, H., Obata, Y., Ohta, N., Inoue, K., Yagi, N., 2010. Chem. Phys. Lipids **163**, 381.

Upgrade of the small angle X-ray scattering beamlines at the Photon Factory

Noriyuki Igarashi[1], Yasushi Watanabe[2], Yuya Shinohara[3], Yoji Inoko[4], Go Matsuba[5], Hiroshi Okuda[6], Takeharu Mori[1] and Kenji Ito[1]

[1] Photon Factory, Structural Biology Research Center, Institute of Materials Structure Science, High Energy Accelerator Research Organization, 1-1 Oho, Tsukuba, Ibaraki 305-0801, Japan
[2]National Food Research Institute, National Agriculture and Food Research Organization, 2-1-12 Kannondai, Tsukuba, Ibaraki 305-8642, Japan
[3] Department of Advanced Materials Science, The University of Tokyo, 5-1-5 Kashiwanoha, Kashiwa, Chiba 277-8561, Japan
[4]Graduate School of Engineering Science, Osaka University, 1-3 Machikaneyama, Toyonaka, Osaka 560-8531, Japan
[5]Department of Engineering, Yamagata University 4-3-16 Jonan, Yonezawa, Yamagata 992-8510, Japan
[6]International Innovation Center, Kyoto University, Sakyo-ku, Kyoto 606-8501, Japan

E-mail: noriyuki.igarashi@kek.jp

Abstract. BL-10C and BL-15A at the Photon Factory, which became operational in 1982, are some of the oldest small angle X-ray scattering beamlines in the world. Recently, both beamlines were upgraded for two-dimensional (2D) SAXS-WAXS experiments. A wide-area imaging plate (IP) detector and a fast-readout flat panel (FP) detector were installed at BL-10C and BL-15A, respectively. Preliminary experiments of both systems showed promising results.

1. Introduction

Small angle X-ray scattering (SAXS) provides information on the structure of matter on the nanometer to micrometer scales. Thus, for years it has been commonplace to investigate complicated hierarchical structures like soft matters and biological systems [1]. The Photon Factory (PF) operates two SAXS beamlines, BL-10C and BL-15A, and partly uses BL-9C for SAXS. Currently, beamtime management and beamline maintenance of BL-10C and BL-15A are performed in cooperation with user groups (Chairpersons are Prof. Yoji Inoko of Osaka Univ. and Prof. Mitsuhiro Hirai of Gunma Univ.). BL-10C and BL-15A were initially installed to provide high X-ray flux for kinetic solution scattering experiments and time resolved muscle fiber diffraction [2]. Both beamlines have been operating for about 30 years, and the last major upgrades were done in the 1990s; thus, much of the hardware and software has since become obsolete.

Soft matter shows a hierarchical structure from the sub-nm to several hundreds μm scale. Hence, the ability to measure scattering over a wide scale is of great importance in the structural analysis of soft matter [3]. To date, we have had many papers on WAXS and SAXS experiments at the both beamlines, but they have been performed separately by assuming the reproducibility of the

Journal of Physics: Conference Series **272** (2011) 012026 doi:10.1088/1742-6596/272/1/012026

experimental conditions. However, irradiation, thermal, or mechanical treatment can easily damage soft or biological samples. Although several users simultaneously performed SAXS-WAXS measurements at the PF, most were limited to one-dimensional measurements using a position-sensitive proportional counter (PSPC). To implement 2D-SAXS-WAXS measurement systems, we recently installed an IP detector system at BL-10C and a FP detector at BL-15A. Herein we describe the development and the preliminary experiment of the 2D-SAXS-WAXS measurement system at each beamline.

2. Wide-area imaging plate system at BL-10C

BL-10C was designed and constructed mainly to investigate non-crystalline materials such as aqueous solutions of biological macromolecules and synthetic polymers. Due to the low sample concentration and low-scattering contrast of such samples, a low-noise signal detection system is required, especially for higher-angle scattering data. For years, only a one-dimensional PSPC (RIGAKU) was used as a detector. However, an IP detector (R-AXIS 7, RIGAKU) equipped with a synchronized X-ray shutter was recently installed to measure wide 2D scattering images (figure. 1).

Preliminary experiments were performed using horse spleen apoferritin and a wavelength (λ) of 0.1488 nm (figure. 2). The camera distance and the exposure time were 1980 mm and 600 sec, respectively. The temperature of the cell, which had a 1 mm beam path and a pair of 20 μm quartz windows, was kept constant at 24°C using a metallic cell holder through which water at a constant temperature was circulated. The standard cell volume was 45 μl (3 × 15 mm^2). Variations in the scattering data for different solutions were corrected by monitoring the beam intensities with an ionization chamber placed in front of the sample cell holder. Two-dimensional scattering data were

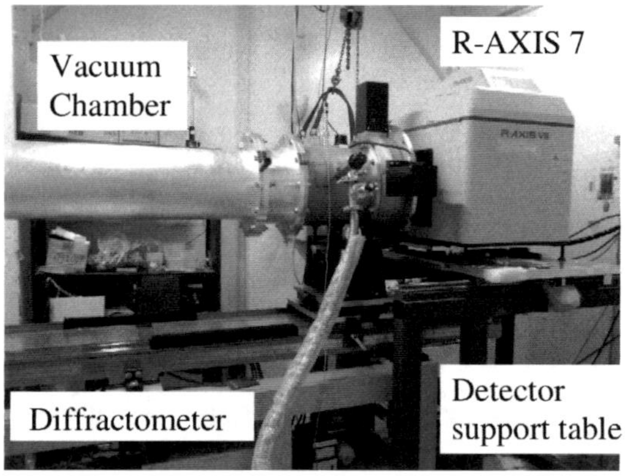

Figure 1. Long camera setting with R-AXIS 7 at BL-10C. The camera distance is about 2000 mm (The middle setting: about 900 mm.). The X-ray beam passes from left to right.

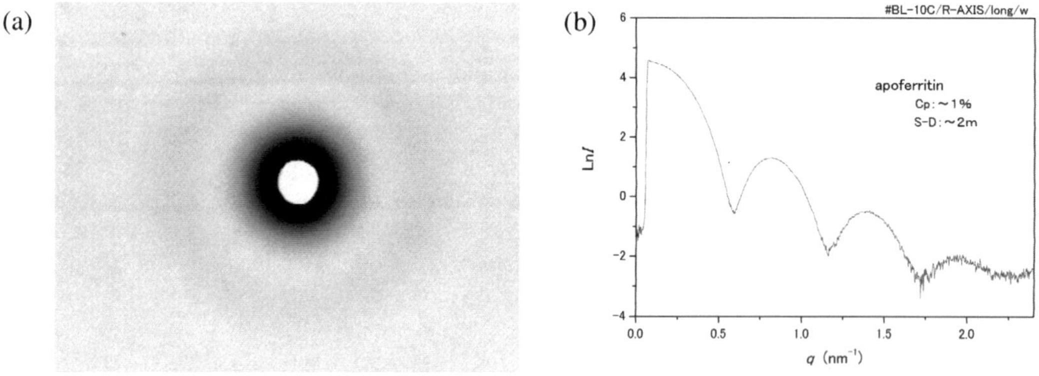

Figure 2. (a) Typical 2D pattern and (b) Scattering curve (*q vs.* ln*I*) of horse spleen apoferritin. Cp is protein concentration and S-D is sample-to-detector distance.

circle-averaged about the beam center. The obtained signals were corrected for both solvent scattering and sample concentration to yield the net scattering intensity *I(q)*. The *q*-value was calibrated by the diffraction pattern of the dried chicken collagen, and the observed *q*-range was between ~0.07 and 2.5 nm^{-1}.

Figure 2b shows a typical scattering curve for the horse spleen apoferritin. Apoferritin is a hollow sphere molecule composed of 24 subunits and each subunit has ~19,800 molecular weight [4]. The scattering pattern indicates peaks at $q = 0.8$, 1.4, and about 1.9 nm^{-1}, and the resolution is sufficient for detailed structural characterization of the 24-mer molecule. This result reveals the benefit of the new IP detector.

3. 2D-SAXS-WAXS system at BL-15A

BL-15A was originally built about 30 years ago for time-resolved studies of small-angle diffraction from contracting muscles [2]. The optics consists of a bent flat mirror and a triangle bent asymmetric-cut monochromator for vertical and horizontal focusing, respectively. It can produce a point focus spot with a high spatial resolution and the intensity necessary for 2D-SAXS-WAXS experiments. Due to such a good point focus beam, the beamline is still widely used for 2D-SAXS and 2D-WAXS. However, the combined measurements of SAXS-WAXS were limited to one-dimensional measurements; thus, we installed a FP detector (Hamamatsu C9728DK-10 [5]) as a WAXS detector for 2D-SAXS-WAXS. Figure 3 depicts the 2D-SAXS-WAXS system. For 2D-SAXS image detection, several detectors are available: two types of CCD detectors (Hamamatsu C4880-10 and C7300) with an X-ray image intensifier [6], pixel array detector (Dectris PILATUS 100K), and off-line IPs (Fujifilm BAS2500).

Figure 4 shows the 2D-SAXS-WAXS patterns for dry chicken collagen. The PILATUS 100K was used for the SAXS detector. The X-ray wavelength and the exposure time were 0.15 nm and 10 sec, respectively. The camera distances of the SAXS and WAXS were approximately 2,000 mm and 50 mm, respectively. Electric pulses generated by an electronic stimulator (Nihon Koden SEN-7103) were used to synchronize the SAXS and WAXS detectors. The SAXS-WAXS images demonstrate that the first order peak is clearly separated from the beam stop and that higher order scattering is observed simultaneously. Although the active area is relatively small, the SAXS-WAXS data can be observed by rotating the sample.

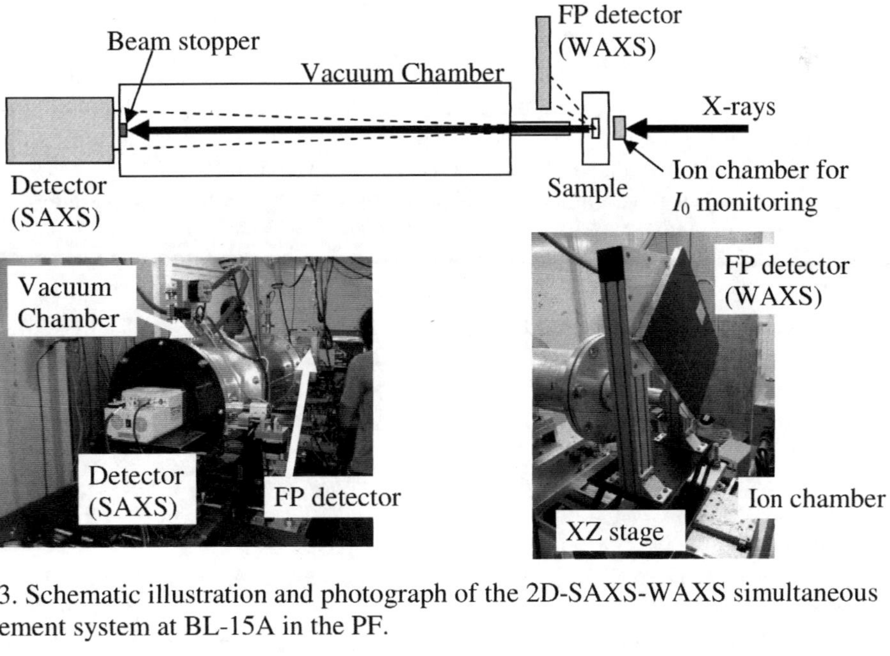

Figure 3. Schematic illustration and photograph of the 2D-SAXS-WAXS simultaneous measurement system at BL-15A in the PF.

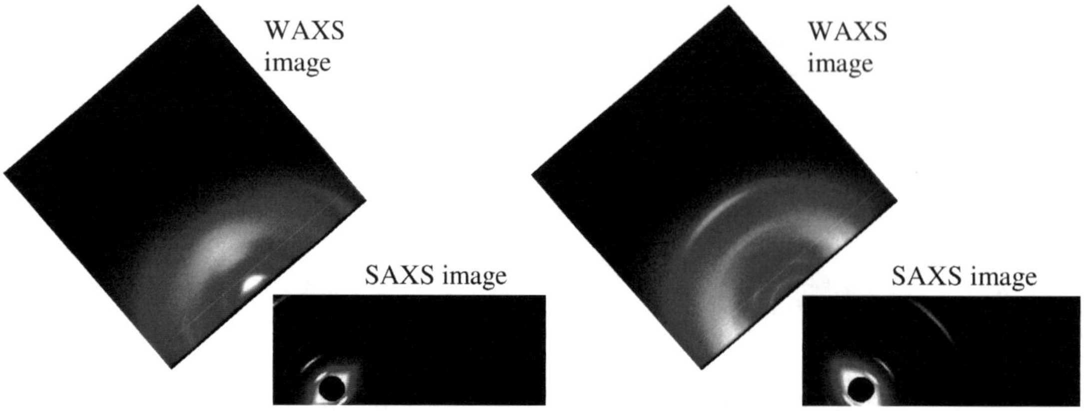

Figure 4. Dry chicken collagen 2D-SAXS-WAXS image. Collagen sample was rotated by 90 degrees around the X-rays axis (right).

4. Further upgrades

Herein we report the successful developments of the 2D-SAXS-WAXS systems at BL-10C and BL-15A at the PF. Additionally, we are pursuing further upgrade programs for both beamlines. The PSPC electronics will be changed from CAMAC to a high-speed data acquisition system based on Si-TCP technology [7]. A GI-SAXS data collection system has been implemented at BL-15A for common use. Moreover, we have proposed a new high-brilliance SAXS beamline at the BL-15 site using a short-gap undulator. This improved BL-15 should be capable of structurally analyzing heterogeneous materials, and then some of the current BL-15A activities will be transferred to BL-6A. BL-6A was closed in the spring of 2010 and the beamline geometry is similar to the current BL-15A. Hence, the current BL-15A settings can be transferred to BL-6A. Then, two bending magnet beamlines (BL-6A and BL-10C) and one undulator beamline (new BL-15A) will be dedicated to the extensive SAXS activities at the PF.

Acknowledgments

The authors acknowledge the other members of the BL-10C and BL-15A operation working group for their long-term commitment to these upgrade programs: Drs. S. Nojima (Tokyo Inst. Tech.), H. Takeshita (Nagaoka Univ. Tech.), H. Kamikubo (Nara Inst. Tech.), Y. Amemiya (Tokyo Univ.), S. Ueno (Hiroshima Univ.), H. Takahashi (Gunma Univ.), Y. Sugimoto (Osaka Univ.), T. Morita (Chiba Univ.), K. Saijo (Kyoto Univ.), K. Yamamoto (Nagoya Inst. Tech.) and M. Hirai (Gunma Univ.).

References

[1] Lindner P and Zemb Th 2002 *Neutrons, X-rays and Light: Scattering Methods Applied to Soft Condensed Matter* (Amsterdam: Elsevier)

[2] Amemiya Y, Wakabayashi K, Hamanaka T, Wakabayashi T, Matsushita T, and Hashizume H 1983 *Nucl. Instrum. Methods* **208** 471

[3] Masunaga H, Sasaki S, Tashiro K, Hanesaka M, Takata M, Inoue K, Ohta N and Yagi N 2007 *Polymer Journal* **39** 1281

[4] Arosio P and Levi S 2002 *Molecular and Cellular Iron Transport* (New York: Marcel Dekker) pp125-154

[5] Yagi N and Inoue K 2007 *J. Appl. Cryst.* **40** s439

[6] Amemiya Y, Ito K, Yagi N, Asano Y, Wakabaayshi K, Ueki T, and Endo T 1995 *Rev. Sci. Instrum.* **66** 2290

[7] Uchida T and Tanaka M 2006 *IEEE Nuclear Science Symposium Conference Record* 1411

Journal of Physics: Conference Series **272** (2011) 012027 doi:10.1088/1742-6596/272/1/012027

Depth distribution of different solvents in a phase-separated block copolymer thin film

N Torikai[1], N L Yamada[2], D Kawaguchi[3], A Takano[3], Y Matsushita[3], E Watkins[4] and J P Majewski[4]

[1] Department of Chemistry for Materials, Graduate School of Engineering, Mie University, Kurimamachiya, Tsu, Mie 514-8507, Japan
[2] Neutron Science Laboratory, Institute of Materials Structure Science, High Energy Accelerator Research Organization, 1-1 Oho, Tsukuba, Ibaraki 305-0801, Japan
[3] Department of Applied Chemistry, Graduate School of Engineering, Nagoya University, Furo-cho, Chikusa, Nagoya 464-8603, Japan
[4] Manuel Lujan Jr Neutron Scattering Center, Los Alamos National Laboratory, Los Alamos, New Mexico 87545, USA

ntorikai@chem.mie-u.ac.jp

Abstract. The structural change induced by the contact with different solvents was investigated for a diblock copolymer thin film, of polystyrene (PS) and poly(2-vinylpyridine) (P2VP), forming a lamellar structure by *in-situ* neutron reflectometry with a conventional cell for the measurement on solid/liquid interfaces. In the case of methanol, which is a selective solvent for P2VP, the structural change was induced by the penetration of methanol into the film, and the original lamellar structure was almost recovered by drying the film even after the solvent contact. On the other hand, when toluene was used as a selective solvent for PS, the block copolymer thin film was easily dissolved into toluene, and its original state was never recovered after the solvent contact. Further, water induced the structural change in the block copolymer thin film similarly to methanol, though it is a poor solvent common for PS and P2VP.

1. Introduction

A block copolymer exhibits rather complicated phase-separation behavior in a thin film, compared with in bulk, due to the interplay of the geometrical confinement effect, the interfacial interaction with a supporting substrate and air, and so on. So far there have been many studies on a block copolymer thin film, and it has been reported on a block copolymer thin film that the orientation of microdomains is dependent on the film thickness, the surface roughness as well as the surface energy of a substrate, etc., and islands or holes are formed on the film surface according to the film thickness relative to the domain size [1]. Also, many attempts have been made to control the self-assembled structure of a block copolymer in a thin film for various practical applications.

Neutron reflectometry is a powerful tool for the structural analysis on material interfaces and thin films because of its high spatial resolution of a few Å in a non-destructive manner. As for the block copolymer thin film forming a lamellar structure, the interfacial thickness between different component lamellae and the segment distribution in a lamellar microdomain were clarified by neutron reflectometry combined with a deuterium-labeling method [2, 3].

In this study, the structural change induced by the contact with different solvents was investigated for the diblock copolymer thin film forming a lamellar structure by *in-situ* neutron reflectometry with a conventional cell for the measurement on solid/liquid interfaces, to understand the interfacial behavior of a block copolymer thin film in contact with a solvent sub-phase.

2. Experimental

The samples used in this study are a hydrogenous polystyrene-*b*-poly(2-vinylpyridine) (*h*-PS-*b*-P2VP) diblock copolymer with the total molecular weight and the volume fraction of a PS block chain in the molecule of about 120×10^3 and 0.50, respectively, and its counterpart (*d*-PS-*b*-P2VP) with almost the same molecular characteristics as *h*-PS-*b*-P2VP except that the PS block chain is deuterated. Both of them exhibit an alternating lamellar structure in bulk. The low molecular weight solvents with different solvency for the block copolymer were used: methanol and toluene as a selective solvent for P2VP and PS, respectively, and water as a common poor solvent for both components of the block copolymer.

The thin film specimens for the reflectivity measurement were prepared by spin-coating 2wt% *p*-dioxane solution of the block copolymer on the polished surface of 10mm-thick quartz blocks with the dimension of 60mm×60mm, and then were annealed at 180°C in a vacuum for 3-5 days before use.

Neutron reflectivity (NR) measurements were performed on two pulsed-neutron reflectometers: SPEAR at Los Alamos National Laboratory, Los Alamos, USA and ARISA-II [4] at the BL16 beamline, Materials and Life Science Experimental Facility (MLF), Japan Proton Accelerator Research Complex (J-PARC), Tokai, Japan. A conventional cell for the measurement on solid/liquid interfaces was used to keep the block copolymer thin film, prepared on the thick quartz block, contact with the solvent sub-phase during NR measurement. The neutron beam is incident from the side face of the quartz block onto the interface between the quartz block and the solvent phase, and the reflected intensity of neutron is detected through the quartz block with a path length of 60 mm. The transmissivity of neutron is estimated to be around 0.2 from a total cross-section for the used quartz. The specular reflection was observed, and the data were analyzed by a model fitting method with the variation in scattering length density along the direction perpendicular to the film surface.

3. Results and Discussion

To observe the effects of solvent contact on the phase-separated structure clearly, most of the following NR measurements were conducted on the block copolymer with a deuterated PS block chain

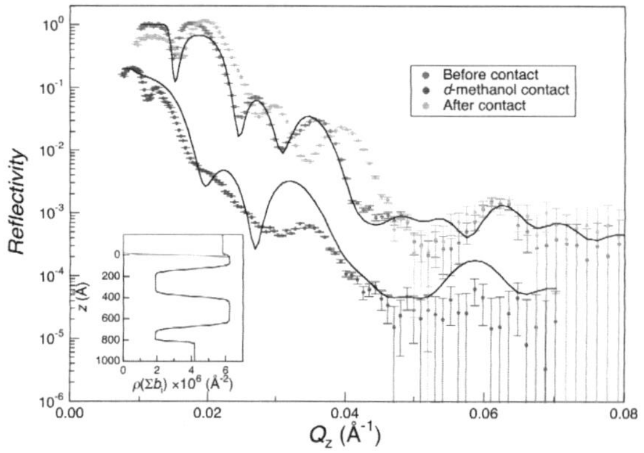

Figure 1. The specular NR profile as a function of Q_z for the *d*-PS-*b*-P2VP thin film in contact with *d*-methanol, along with those before and after the solvent contact.

unless it is mentioned. Figure 1 compares the specular NR profile for the *d*-PS-*b*-P2VP thin film in contact with deuterated methanol (CD_3OD), which is a selective solvent for P2VP, with those taken in a dried state before and after the solvent contact, as a function of neutron momentum transfer, Q_z ($=(4\pi/\lambda)\sin\theta$, where λ and θ are the wavelength and the incident angle of neutron, respectively), along the direction perpendicular to the film surface. It should be noted that the reflectivity profiles for the thin film in the dry state were measured by a normal manner, in which the neutron is incident from air onto the film surface prepared on the quartz block. The reflectivity profile (red closed circles) for the film before the solvent contact presented a few distinct Bragg peaks implying that the formed lamellar microdomains are preferentially oriented along the direction parallel to the film surface. The solid black line superimposed on the experimental data is the preliminary fitting result calculated from the variation in scattering length density, $\rho(\sum b_i)$, as shown by the red line in the inset, along the perpendicular direction to the film surface. This indicates that *d*-PS and P2VP lamellae are alternatively stacked with the domain spacing of about 530 Å in the 800Å-thick film. It was also found that *d*-PS lamella appears at the air surface of the film due to its lower surface energy, while P2VP lamella does at the interface of the film with the quartz block because of its preferential affinity to the quartz.

When this *d*-PS-*b*-P2VP thin film was made contact with *d*-methanol, which is a good solvent for P2VP but a poor solvent for PS, the Bragg peaks in the reflectivity profile (blue closed circles) became weaker than those before the solvent contact. It should be noted that the critical Q_z position for total reflection was shifted to the lower-Q_z region, since the neutron reflection was observed through the quartz block with relatively high $\rho(\sum b_i)$ against the deuterated solvent, though the transmissivity of the quartz block is not exactly considered into normalizing the reflectivity data. The black solid line on the measured profile was calculated from $\rho(\sum b_i)$ variation, indicated by the blue solid line in the inset, assuming that *d*-methanol does not penetrate inside the film at all so that no structural changes are induced. It is apparent that the reflectivity profile thus calculated does not agree with the experimental data. This implies that some amount of methanol penetrates into the film, and induces the structural change of the lamellar microdomains during the contact with solvent. Levicky et al. [5] previously reported similar results by a NR measurement that methanol induced the selective swelling of P2VP lamellae for a PS-*b*-P2VP diblock copolymer thin film. Then, by drying the film after the solvent contact, the Bragg peaks in the profile (orange closed circles) were almost recovered but their position was shifted to the higher-Q_z region, reflecting that the size of lamellar microdomains became smaller. This may be attributed to a kind of solvent-annealing effect by the penetration of methanol into the film during the contact.

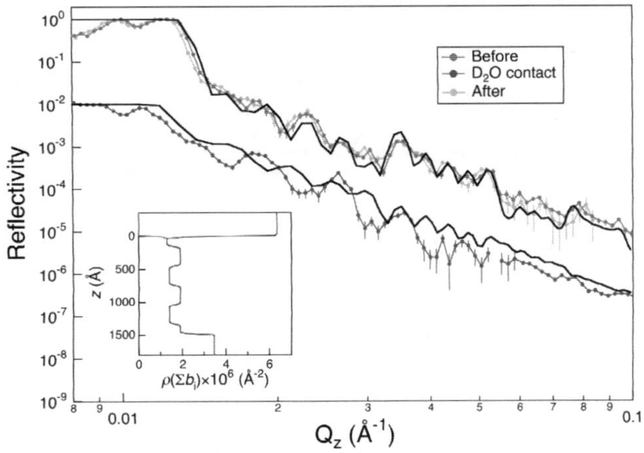

Figure 2. The specular NR profile for the *h*-PS-*b*-P2VP thin film in contact with D_2O, along with those in the dried state before and after the solvent contact.

In the case of *d*-toluene, which is recognized as a good solvent for PS but a poor solvent for P2VP, the distinct Bragg peaks observed in the dried state almost completely disappeared by the contact with toluene, though the reflectivity profiles are not shown here [6]. Those Bragg peaks were never recovered even by drying the film after the solvent contact. This suggests that the block copolymer thin film was easily dissolved into toluene, compared with methanol described above. This difference between toluene and methanol may be caused by their subtle difference in the solvency for the component polymers, though both are a selective solvent.

When the *d*-PS-*b*-P2VP thin film was made contact with deuterated water, D_2O, which is known as a common poor solvent for PS and P2VP, the induced change in the reflectivity profile of the block copolymer thin film was similar to the case of methanol, though the reflectivity profiles are not shown here [6]. However, the sample labeling adopted was not suitable for examining whether water penetrates into the film, since both the block copolymer and the solvent are deuterated. Thus, to facilitate observing the distribution of solvent in the block copolymer thin film, non-deuterated *h*-PS-*b*-P2VP thin film was prepared and made contact with deuterated solvent, D_2O. As presented in Figure 2, the reflectivity profile (red closed circles) for the non-deuterated block copolymer film before the solvent contact exhibits the weak peaks originated from the phase-separated structure, since the difference in $\rho(\sum b_i)$ is not so large between a non-deuterated PS and a P2VP. The black solid line on that profile is the reflectivity calculated from $\rho(\sum b_i)$ distribution, shown by the red solid line in the inset, along the direction perpendicular to the film surface suggesting a phase-separated state in the film with the thickness of about 1,500 Å although the calculated reflectivity is very similar to but not perfectly fitted with the experimental one yet. As this film was made contact with D_2O, the reflectivity profile presents slightly more discernible peaks in a low-Q_z region than that before D_2O contact. Actually, the calculated reflectivity profile does not agree with the experimental data (blue closed circles), assuming no penetration of D_2O into the film so that simply replacing an air phase with a D_2O sub-phase without any structural changes of the thin film, as shown by the blue solid line in the inset. This implies that some D_2O may penetrate into the film somehow though it is a common poor solvent for both the components, that is, PS and P2VP, of the block copolymer. At present, the exact mechanism for the water penetration is not clear yet. Further experiments as well as data analysis are required to clarify the mechanism of water penetration into the block copolymer thin film.

Acknowledgements

The recent NR measurements using the non-deuterated block copolymer to examine the depth distribution of solvent were performed under the S-type research project, KEK (2009S08), and the others were done under the Inter-University Research Program on Pulsed-Neutron Scattering at Oversea Facilities. N. T. is indebted to the supports by a Grant-in-Aid for Creative Scientific Research (16GS0417) from the Ministry of Education, Culture, Sports, Science and Technology of Japan.

References

[1] Hamley I W 2009 *Prog. Polymer Sci.* **34** 1161-210
[2] Torikai N, Noda I, Karim A, Satija S K, Han C C, Matsushita Y and Kawakatsu T 1997 *Macromolecules* **30** 2907-14
[3] Noro A, Okuda M, Odamaki F, Kawaguchi D, Torikai N, Takano A and Matsushita Y 2006 *Macromolecules* **39** 7654-61
[4] Mitamura K, Yamada, N L, Sagehashi H, Seto H, Torikai N, Sugita T, Furusaka M and Takahara A *J. Phys.: Conf. Ser.* submitted
[5] Levicky R, Koneripalli N, Tirrell M, Ankner J F, Kaiser H and Satija S K 1998 *Macromolecules* **31** 4908-14
[6] Torikai N, Yamada N L, Kawaguchi D, Takano A, Matsushita Y, Watkins E, Majewski J P and Okuda H 2007 *J. Phys.: Conf. Ser.* **83** 012028

AUTHOR INDEX

AUTHOR INDEX